SCIENCE SCOPE

by
Kathryn Stout, B.S.Ed., M.Ed.

A Guide for Teaching Science in Grades K - 12

Including
Concepts
Objectives
Teaching Tips

A DESIGN-A-STUDY BOOK

OTHER TITLES BY KATHRYN STOUT:
Comprehensive Composition
Critical Conditioning
Guides to History Plus
Maximum Math
The Maya
Natural Speller
Teaching Tips and Techniques

Audiocassettes:
How To Teach Composition
Make It Easy on Yourself
Math That Makes Sense
Teaching Kids to Think
Teaching Reading, Spelling, & Critical Thinking
Teaching Tips That Really Work

Current listings and prices available from Design-A-Study at
Web Site: http://www.designastudy.com/
E-mail: kathryn@designastudy.com
Phone/Fax: (302) 998-3889
or write to the address below.

Published by Design-A-Study
408 Victoria Avenue
Wilmington, DE 19804-2124

Cover Design by Ted Karwowski and Richard B. Stout

ISBN 1-891975-03-X

Library of Congress Catalog Card Number: 98-92682

CONTENTS

Introduction

Science Scope provides lists of skills, topics, and concepts that are part of most science curricula. It can be used:

 As a checklist.

This allows the student to use a wide variety of materials to learn about science instead of a program designed by only one company for use in kindergarten through high school. Teachers and parents can also be sure a depth of understanding is achieved, rather than simply exposure to a topic. Older students can use this book to help them recognize what they should understand and retain (checking off items themselves) whether using a text book or several books and resources.

 As an overview of the progression of a topic from simple concepts to the more complicated.

Even if a topic is not introduced until a student is eleven or twelve, for example, it may be helpful to introduce it using ideas from the easier level. This guide includes material used in high school courses for the college-bound student, which may prove too complex for some students. In that case, teachers and parents have the option of taking the student as far as he is able in any topic, while making certain that basic concepts are understood.

 As a reference for age-level expectations.

Find out how much can be expected at any age. Keep children challenged, but not beyond their ability to meet that challenge. Topics are divided according to the following key:

Primary	Ages 4 - 8	Grades Pre K - 3
Intermediate	Ages 9 - 11	Grades 4 - 6
Junior High	Ages 12 - 14	Grades 7 - 8
High School	Ages 15 - 18	Grades 9 - 12

The age ranges are provided as a guide, not as a requirement for completing, or even introducing, topics at that age.

TEACHING SCIENCE

How do scientists come up with a concept to pass on?

First, someone asks a question. Wanting to understand why things happen as they do leads to a search for answers. The search leads to observations, experiments, and the organization of all the collected information in order to come up with a logical conclusion. When new information is discovered, the conclusion must be looked at again. If the new knowledge gives support, fine, but if not, what looked like an answer must be discarded or modified. That means scientific ideas may change. The changing attitudes and practices in medicine throughout history illustrate that science is a process.

Searching for answers may be done fairly informally. But, at some point, all the observations must be organized in such a way that the conclusions drawn can be tested by someone else with the same result. This is called the "scientific method." A scientific "law" is the conclusion or generalization that represents all the known facts. It is the law that changes with new findings. Children must learn to use the scientific method, not just define it.

WHERE TO START

 Choose a goal. What understanding do you want to develop?

 Choose one or more activities that will allow students to use some of the skills listed in *Scientific Skills*.

 Encourage discovery by asking questions rather than offering explanations. Let students look for the answers. This also encourages logical thinking.

How many (legs) does it have? *(Instead of "A bird has two legs.")*

What do you think will happen if _____?

Why do you think _____?

What could you change so that _____ would (happen)?

How would (our life, its life) be different if . . .
("...there were no gravity." "...it didn't have any predators.")

Activities should give the students the opportunity to do at least some of the following:

▶ Describe or explain.

▶ Recall
  • Information already gained that may apply.
  • Conditions of an experiment from the previous day.
  • The order in which events occurred: What happened first? Next?

▶ Speculate or predict.

▶ Collect and organize his own data and draw his own conclusions.

▶ Ask questions that can lead to further study.

 ## Guide students toward discovery.

Suggest questions, and sometimes offer information that students could not discover (in the situation) in order to guide students toward discovering some central theme or point. This process should include helping students decide which factors are relevant (useful), and which are irrelevant.

Skills that involve reasoning include: inferring, predicting, controlling variables, interpreting data, and/or formulating a hypothesis.

 ## Teach the student about variables and the need for a control in experimentation.

Introduce the idea of variables and the need for a control when questioning leads to conducting an experiment in order to find answers, or to test ideas. This can be introduced by questioning. For example, if trying to determine the cause in a change in the weight of a hamster, ask "If we give him different food <u>and</u> different exercise equipment, how will we know which caused any changes we see?" Lead the student toward the understanding that only one condition can be changed at a time.

 ## Teach students not to jump to conclusions.

When you want to help students realize that they must not jump to conclusions based on one experiment, perform one experiment several times to see if the results are the same, or conduct several different experiments to prove one point.

Follow up

Students can easily get lost in the process and never really understand the point (your objective). Therefore, it is essential that you conduct some kind of follow-up. Here are a few types of follow-up to choose from, or you may think of other ideas:

▶ **Discussion**
Discuss the experiment and results (review and summarize) and what was learned in order to prevent mistakes from being repeated, or to re-emphasize the point.

▶ **Application**
Have students draw on what they've learned in order to deal with a current situation. For example, let a young child check the weather and then choose his own clothing after a lesson on how weather determines how people dress.

▶ **Drill**
Construct and administer some sort of test (written or oral) or contest.

 **TEACHING TIPS**

- Experiments develop important skills. (See *Scientific Skills*.)

- Demonstrations illustrate a generalization.

You must decide what the focus should be and then determine the most appropriate type of activity. *A number of different types of activities can all be the result of searching for answers to questions:*

> ▶ Conduct an experiment.
> ▶ Make observations.
> ▶ Interview people.
> ▶ Study books written on the subject.
> ▶ Watch someone conduct experiments and/or listen to explanations of a concept.

As a teacher you can:

- Plan field trips to the zoo, park, science museum, pond, woods, ocean, gardens, supermarket, and so on.

- Videotape television programs dealing with science topics.

- Provide kits with equipment and explanations for experiments children can do at home. (Or provide information for experiments that only require materials easily available in the home.)

- Use diagrams or other visual aids while children discuss information they would not otherwise be able to observe—the internal body organs for instance.

- Read aloud biographies of scientists that will stimulate students to discover on their own as well as help them understand the scientific process at work.

- Provide follow-up to an activity so that the generalization (your goal) is recognized or reaffirmed.

☞ *It would be impossible to structure every learning situation so that a child discovers the information you want him to learn. However, keep the following ideas in mind as you plan:*

1. The younger the child, the more he needs to involve his senses ("hands-on") in order to develop a real and lasting understanding. Memorization has its place, but is not a useful tool for developing comprehension.

Maturity does bring a greater ability to learn by reading without depending on "doing" for understanding. This ability is enhanced by a background of experiences and critical thinking.

2. The younger the child, the more discovering on his own makes an idea exciting, valuable, and worth remembering.

With maturity a student becomes more able to discover by asking questions and looking for answers in material as he reads. This approach to reading can be taught at an early age so that it becomes a natural approach rather than the typical habit of memorizing, and thereby accepting, everything read.
(The Design-A-Study guide *Critical Conditioning* explains this approach.)

3. Experience, not memorized information, provides a foundation for later comprehension of more complicated information.

Experience can include doing, discovering, dramatizing, or discussing an idea.

For example, a child may be able to solve pages of math problems with complete accuracy, but still not know what to do when faced with a real-life situation that requires him to decide which type of math to use. If he understands the concepts of adding, subtracting, and so on, he will automatically apply that understanding while in the grocery store, or while figuring out how to earn enough money to buy something he wants.

4. The younger the child, the more concrete (see it, hear it, touch it) the experiences and concepts should be. As children mature, they are able to handle more abstract ideas.

5. Drill is useful for developing quick recall of information, but is not appropriate for initial learning unless it's not necessary to develop any understanding of the information.

> **The younger the child, the more activities should favor field trips, experiments, and investigations.**

Older students should continue to have these experiences, but can handle more reading and memorizing of information already discovered. Always, though, their minds should be asking questions about what they read. Allow them to question its truth: Who said it? When? Based on what evidence?

Older students should be given opportunities to apply what they are learning.

For example, students should be allowed to question texts that offer evolution as a fact, not a theory. Doubts about "facts" offered should lead to questions and a search for more information from other scientists that continues until the students come to a conclusion based on logic and evidence, even though they have not conducted any direct observations or experiments. They could apply knowledge gained by representing the "con" side in a debate, or by writing a composition with adequate and detailed support for their position.

SCIENTIFIC SKILLS

If students are to think and behave scientifically, they must:

- Learn to make observations. (This becomes their data, or information.)

- Find some pattern or order in what they observe in order to form an hypothesis.

- Test that hypothesis with an experiment.

These can be done at every level, primary through high school. Begin with simple ideas and experiments and gradually move toward the more complicated. Incorporate scientific skills at every age. Primary students should be specifically taught the first five skills listed below. Older students (grades four and up) can also handle the last six.

OBSERVING

Using one or more of the senses, identify and name properties of whatever is being observed. Work toward accuracy and detail.

Ask   What do you See?
        Hear?
        Feel? ("Feel" by touch, not "feel" as an emotion.)

At first, children may confuse their own inferences with observations. Observations provide the basis for an inference or hypothesis, but must be recognized as separate.

For example, when asked for an observation, a child might say, "It's a ball." That would not be a property and, therefore, should not be listed with observations. It is a conclusion reached after looking at some of the properties—it's round and bounces.

▸ Use as many of the five senses as possible.
    Hear - Sounds
    See - Size, Shape, Color, State (solid, liquid, gas)
    Smell - Any odor
    Taste - Bitter, Salty, Sour, Sweet
    Touch - Texture, General Temperature (warm, cold)

▶ Be specific in describing what is observed.
  Aqua is more specific than *blue*.
  Makes a bell-like sound is better than *rings*.
  Three leaves rather than *many leaves.*

▶ State the observations in quantitative terms.
  Give a standard unit of measurement: size, weight, temperature
    ⇒ *A big bug* is not as accurate as, *A bug 1 inch long and 1/4 inch wide.*

    ⇒ Young children could compare the size to some other object until they are able to measure—"As big as a penny."

▶ State any changes observed.
    ⇒ This might involve doing something to the object: hitting it with a mallet, heating it, putting it in water, and so on, and then stating an observation of whatever changes result.

    ⇒ State how long it took before seeing a change in seconds, minutes, hours, or days, if applicable.

▶ Use tools.
  ◆ Magnifying glass
  ◆ Measuring devices—ruler, meter stick, thermometer
  ◆ Microscope
  ◆ Scales of various types
  ◆ Telescope

CLASSIFYING

Sort a collection of objects or events in an order based on some observable property or properties.

This skill is commonly referred to in primary activity books as *grouping,* and is followed by questions encouraging children to look for ways things are alike and different. Classification activities are also part of most math curricula. You can use prepackaged materials with ideas for comparing, contrasting, and finding interrelationships, or use an informal approach with objects around the house. The point is to lead children toward an understanding that we group according to characteristics we choose simply because it is useful for some purpose. Therefore, there is no one correct way to classify; correctness has to do with being certain that each item in a group has the same property in common.

1. Begin with dividing objects into two or more subsets based on one observable property:
    - ◆ Color
    - ◆ Shape
    - ◆ Size
    - ◆ Taste
    - ◆ Texture
    - ◆ Use or function

2. Divide objects within each group further according to a second observable property.

3. Continue breaking down subsets into smaller categories.

4. Develop a classification key to state how the objects are organized. (See the model that follows.)

This skill aids later understanding of the classification system used to separate all living things into kingdoms, phyla, classes, and so on. It also helps children recognize the usefulness of classifying elements as they appear in the periodic table.

Groupings made from a pile of blocks could be recorded using a classification key like the one below:

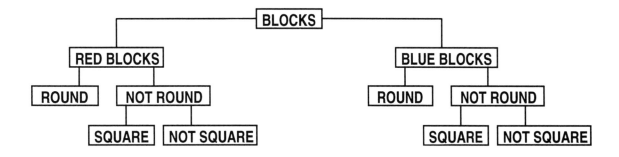

Ask What belongs together?
Based on what criteria?

How would you name these groups?

What belongs under what?

MEASURING

Science often requires precise and accurate measurements.

▸ Use a variety of instruments and standard units.

    Instruments: ruler, scale, thermometer (etc.)
    Standard Units: inches/centimeters; ounces/grams; degrees (etc.)

Become familiar with the metric system since that is the system of measurement used by scientists internationally.

▸ Decide which instrument to use.

▸ Carry out calculations.

▸ Work toward precision and accuracy.

The smaller the unit of measurement used, the more precise the measurement. For example, a line may be measured to the nearest centimeter or to the nearest millimeter. The endpoint of the line is likely to fall between markings, and would be closer to, if not at, a millimeter marking than the marking for a centimeter.

In comparing several measurements of the same object, the most accurate would be the measurement using the greater number of significant digits. For example, 4.053 cm would be more accurate than 4.0 cm.

Differences in results in measurement can be the result of different people doing the measuring, or the use of different instruments. Therefore, some error in measured value is to be expected. If a group of students each measures the same object, recording varying results, the mean of their measurements should be used. (They simply add their results and divide by the number of measurements to arrive at the mean, or average. For instance measurements of 2.5 mm, 2.6mm and 2.5 mm would be added, and the total, 7.6, divided by 3, giving a mean of 2.533.)

Math programs usually provide instruction and practice in measuring as well as computation. Science can give students an opportunity to see how math can be useful.

At the junior and senior high level, measurements will include an understanding of area, mass, volume, density, power, and force. These are noted in this guide when appropriate.

PREDICTING

State a prediction based on observed events by

interpolation *(Predicting a value between or within two given values.)*
or
extrapolation *(Seeing a pattern and predicting the next, or future, value.)*

▶ Make or use tables or graphs of data to make predictions.

▶ List predictions according to your confidence in them.

Confidence should be determined by how many variables are constant, and how many are unknown, or not constant.

Ask   What would happen if . . . ?
          Why do you think that would happen?
          What would it take for such-and-such to be true or probably true?

Math studies in probability help develop this skill.

Applying science involves predicting: "If I want a light to go on, I should attach this wire." This is where science becomes practical.

INFERRING

An inference is an explanation of an observation.

▶ State an inference from a set of observations.

▶ Identify the observations that support the inference.

▶ State additional observations that are needed to test alternative inferences.

▶ Identify inferences that should be accepted, rejected or modified based on additional observations.

Ask   Why did such-and-such happen?
        What does this mean?
        What would you conclude?

This is a fundamental part of logical thought and of science. The inference is what makes the observations useful.

Students should be encouraged to state more than one inference to explain the observation or set of observations. They should then look at what would be needed in order to come to a more certain conclusion. This can also have the benefit of helping them learn not to jump to conclusions in daily life.

Begin practicing the following skills in intermediate grades while continuing with those already listed.

CONTROLLING VARIABLES

Variables are factors that influence behavior or properties. Being able to identify variables and then control them allows us to come to more reliable conclusions.

▸ Identify variables.

▸ Only manipulate one variable at a time, leaving the others constant.

▸ After manipulating one variable, watch for and measure changes that take place in response.

For example:
If you wanted to find out which of a list of ingredients gave the cookie its sweet flavor, you would make the recipe again leaving out only one ingredient. You would repeat this, never leaving out more than one ingredient, until when tasting the cookie, the sweet taste was missing. Then you would record that leaving out "sugar" (or whichever sweetener was left out) resulted in a cookie without a sweet flavor.

INTERPRETING DATA

Summarize the data and construct one or more predictions, inferences, or hypotheses, from the information.

Facts alone are of little value. What we do with those facts requires an interpretation—forming an opinion or drawing a conclusion. While this is part of the scientific method, it is such an integral part of our daily lives that it is also taught in reading and math programs and in logic workbooks.

The interpretation of the experiment described to find out what gives a cookie the sweet taste could be any of the following:

♦ When sugar was left out of the recipe, the cookie did not taste sweet. If we try the recipe again without sugar, the cookie won't taste sweet. (*Prediction*)

♦ Sugar gives the cookie it's sweet taste. (*Interpretation*)

♦ Recipes that do not use sugar won't taste sweet. (*Hypothesis*)

The hypothesis could then be tested to prove it false. (It is a conclusion a child may draw and want to test.)

FORMULATING AN HYPOTHESIS

This is a generalization of observations or of inferences—an explanation of why things happened as they did. It is an answer broad enough to be applied to other, untested situations.

Students will draw on past experiences and knowledge when formulating an hypothesis.

▶ Once a generalization is suggested, a question must be asked:

"Does the idea work when it is tried?"

And so, while this is not listed as skill number one, this is the first step.

▸ A test is constructed, predictions are made, and observations, experiments, and investigations are carried out.

The hypothesis that not using sugar would result in no sweet taste is a generalization. The result of one experiment is generalized to apply to other untested situations. Experiments using honey, molasses, fruit juices, or artificial sweeteners would prove it to be a false generalization, and the hypothesis would have to be discarded.

If you mixed sugar in water, using cold and then hot water, you could hypothesize that substances will dissolve faster in hot water than in cold.

After testing several other soluble solids, you would either revise the hypothesis, discard it altogether, or stand behind it as provable. (If there are a few exceptions, the modification might be: Most soluble substances dissolve faster in hot water.)

DEFINING OPERATIONALLY

Whenever possible, a definition should include what you do and what you observe when experimenting in the physical sciences.

What those things are will depend on the situation. Usefulness is the key. There is no one correct operational definition for every experiment. The definition needed depends on such things as the ages of the students and what is being learned.

Using an operational definition simply means making yourself clear and avoiding vague language while explaining what to do and what to look for so that others can copy your experiment, observe what you observed, and reach the same conclusions.

Biology as a science tends to rely on descriptions. However, when experimenting in other areas, provide an operational definition for the young or inexperienced and gradually help them develop their own.

For example, if you are studying electricity, an operational definition of a conductor might be, "Any object which can be used to complete an electric circuit." That would give others something *to do* and *to observe* in order to decide whether or not the object tested can be classified as a "Good conductor of electricity."

For children who don't understand the term "completing a circuit," a more useful operational definition would be, "If a light bulb is glowing, the object is used to replace a wire and if the light goes on again, the object is a good conductor of electricity."

Another example would be to describe rain simply as drops of water falling from a cloud in the sky when teaching the young. At the high school level, students look more thoroughly into the water cycle and use more precise language in describing rain from a scientific point of view.

Following are two examples of definitions and operational definitions to illustrate the difference:

1. Equal masses: Two objects that weigh* the same.   *Definition (*See page 104.)*

    Two objects that balance each other on an equal-arm balance.
    Operational definition. It implies what to do and what to observe.

2. Ice: A solid made up of water molecules.   *Definition*

    A solid that changes to a liquid at 0 degrees Centigrade.
    Operational definition: It implies what to do and what to observe.

EXPERIMENTING

Experimenting makes use of all the scientific skills.

1. Define the problem.
    ⇒ Is a hypothesis being tested?
    ⇒ Is a question to be answered?

2. Design the experiment or test.
    ⇒ What will you do?
    ⇒ How will it be done?
    ⇒ What information are you looking for?
    ⇒ What are possible problems that you could encounter?
    ⇒ How can they be dealt with?

    This leads to:
    ▸ Identifying the variables that are to be controlled.
    ▸ Constructing any necessary operational definitions.
    ▸ Making predictions of possible outcomes.

3. Carry out the test, make observations.

4. Collect and interpret the data.

5. Report the test and the findings in enough detail that someone else can repeat it exactly and get the same results. Include a statement as to whether or not the hypothesis is supported.

Coming to a conclusion often leads to other questions:

Will the hypothesis be supported every time?

If the hypothesis is unsupported: "Could the reason be ___ ?" This leads to a new hypothesis, requiring further experimentation.

RESEARCHING

Students should not see their own efforts as the sole source of knowledge. They should find out what others have done and how those results compare to one another and to their own results.

They can accomplish this in various ways.

▸ Checking with other students, and teachers.

▸ Watching demonstrations by someone in the field.

▸ Using library resources to find out about other experiments and their results.
   ◆ Accessing information via the Internet
   ◆ Biographies of scientists
   ◆ Books on the subject that may include references to tests done
   ◆ Periodicals
   ◆ Reference books

LIFE SCIENCE

ANIMALS

Categories: Living and Nonliving

Things can be divided into two categories: living and nonliving.

PRIMARY

BEFORE separating plants and animals, children should classify things as living or
nonliving. They could sort pictures or point to objects. Then have them offer reasons for
putting things into one category or the other.
Living: breathes, grows   Nonliving: doesn't breath, doesn't grow

INTERMEDIATE

Students begin to look more closely at ways to separate living from nonliving. Introduce
any (or all) of the characteristics listed below.

JUNIOR HIGH - SENIOR HIGH

Processes or Activities characteristic of living things:

1. **Movement** *(Growth)*: Although not every living thing can move from place to place,
   there is movement in growth (by increasing cell size or by adding new cells) or in the
   flowing of material within a cell (in one-celled organisms, for example).

2. **Use of Food** *(Nutrition)*: Food is used for growth and for the release of energy.

3. **Respiration**: Taking in something and giving off something else. (This involves
   chemical changes.) Most animals breathe in oxygen and breathe out carbon dioxide.
   Oxygen is used to burn the food and release energy.

4. **Reproduction**: Each living thing must be able to make another organism like itself.
   One-celled organisms divide into two complete cells.

5. **Response to Stimuli**: Living things (even one-celled organisms) react to changes
   within their surroundings. (*heat, light, sound, chemicals*)

---

6. **Repair** *(Regeneration):* At least some parts of an organism can be replaced if destroyed. (Observe the healing of a cut.)

Some texts also include:

7. **Metabolism**: Various chemical and physical changes occur continually in living organisms. (This covers how food is used once it is taken in.)

Students should be able to define: respiration, excretion, metabolism, nutrition, asexual and sexual reproduction, stimulus-response.

Other characteristics of living things:

1. **Living things are made of PROTOPLASM.** Only living things can repair damaged protoplasm or make new protoplasm.

Cells are the building blocks of living things. Cells contain protoplasma—a colorless (usually), jelly-like substance made up of carbon, hydrogen, nitrogen, oxygen, water, and sometimes sulfur and phosphorus.

2. **Living things are ORGANIC.** That is, they contain carbon.

All MATTER is classified as ORGANIC or INORGANIC. Organic means it contains carbon; inorganic means it does not contain carbon. Don't let students become confused by this. Although most things that are organic are living, there are nonliving things that are organic as well. For example, coal contains carbon, but, obviously is not alive. This characteristic becomes more clear if presented along with the carbon cycle.

ANIMALS: Distinguishing Characteristics

Plants and animals are both living things, but with characteristics that allow us to tell them apart.

PRIMARY
Have children classify objects as plant or animal and explain why they put them in each category. *(Animals can move place to place; plants move only by growing.)*

INTERMEDIATE - JUNIOR HIGH
A closer look at cells and at plants will help students discover that:

▸ Only plant cells have a cell wall.

▸ Only plants have chlorophyll and can make their own food (photosynthesis).

▸ Only animals have a nervous system.

ANIMALS: Classifications

**Animals have definite physical characteristics that allow us
to distinguish one from another and put them into groups.**

PRIMARY

Use observation and classification skills.

▸ Describe ways to tell one animal from another.

> *"How is this bird different from this cat?"*
> *"How are they the same?"*

▸ Put animals into groups and tell why you organized them the way you did.

Possible groups:

By Size: Big animals
                               Small animals
(The child compares the animal to himself to determine size.)

By Use: Pets
                              Zoo animals
                              Circus animals
                              Farm animals
(Typical groups in picture books.)

By General Habitat: Birds *(Air)*
                              Fish *(Water)*
                              Land animals

By Specific Habitat: *Desert* animals
                              Ocean animals
                              Polar animals

▶ Describe ways in which animals are alike and different within a group, and between groups. *Include*:

♦ **Size**
⟹ Big or small compared to what?
(Help young children realize that they use "big" or "small" in comparison to themselves and that size is relative.)

♦ **Body parts**
⟹ How many
⟹ Size
⟹ Shape

♦ **Body covering**
⟹ Feathers
⟹ Fur
⟹ Hair
⟹ Shell
⟹ Skin

♦ **How the animal moves** *(It may move in several ways.)*
⟹ Crawl
⟹ Fly
⟹ Hop
⟹ Swim
⟹ Walk

♦ **How the animal breathes**
⟹ Gills
⟹ Lungs

▶ **Given the following categories, observe examples and discover characteristics that belong to each group:**

Category	Characteristics
Birds	two legs, two wings, covered with feathers
Fish	backbone, gills, fins
Mammals	body hair, babies get milk from their mother's bodies
Reptiles	cold-blooded
Amphibians	live part of their lives in water (as babies) and part on land
Insects	six legs, two antennae, three body-parts
Arachnids	eight legs, no antennae, two body parts

At least one animal in each group should be studied more thoroughly.

Ask

1. What does it look like at birth? How do its looks change as it grows?
2. What does it eat? What eats it?
3. What does it need to survive?
4. How do the parents care for the babies?
5. Where do they live (habitat) and why?
6. How do they adapt to their environment?

> The most common amphibians studied are frogs, toads, and salamanders. Children enjoy catching tadpoles and watching them develop.

INTERMEDIATE - JUNIOR HIGH

Students become acquainted with more groups of animals and their scientific classification. Use the chart of *Vertebrates and Invertebrates* for quick reference.

1. Give students classifications and their descriptions and have them sort animals into the appropriate category.

or

2. Give students a number of animals in each category and the classifications (birds, fish, mammals. . . .). Then have them look for common characteristics in order to figure out the basis for each category on their own.

Classification of living things continues to change. Early texts divide everything into either the Animal or the Plant Kingdom. Later, a third kingdom was introduced called the Protist Kingdom. The protozoa listed in the Animal Kingdom was reclassified as a Protist, along with several plants: algae, fungi, lichen, slime molds, and bacteria.

Most recently, five divisions are in use: **Animal, Plant, Protist, Fungi**, and **Monera Kingdoms.**

Monera Kingdom	Single-celled organisms without a nucleus. (The nuclear material is scattered.) *bacteria, blue-green bacteria*
Fungi Kingdom	Many-celled organisms that cannot make their own food (not green). *mushroom, mold, yeast*
Protist Kingdom	Single-celled organisms with a nucleus. *protozoa (paramecium, amoebae), slime mold, lichen*

Plant Kingdom   Single or many-celled organisms that make their own food, and cannot move from place to place. *algae, fern, tree, flowering plants*

Animal Kingdom   Many-celled organisms that cannot make their own food and can move from place to place.

Within each Kingdom are groups called **phyla**. Each phylum is divided into classes. At the high school level, students should become aware of the further divisions: **order, family, genus,** and **species.**

At this level have students group animals as vertebrates or invertebrates. Later, explain that those listed as vertebrates are in the phylum Chordata, and each category is actually a class, while those listed under invertebrates are all phyla.

There are nine (or ten) phyla in the Animal Kingdom:

1. Chordata (all the vertebrates)

2. Arthropods

3. Coelenterates

4. Echinoderms

5. Mollusks

6. Porifera

Worms
7. Platyhelminthes (flatworms)

8. Nematoda (roundworms)

9. Annelida (segmented worms)

And, if using only two kingdom divisions:
10. Protozoa (single-celled) (otherwise, part of the Protist Kingdom)

In the **phyla Chordata**, there are five **classes**:
1. Amphibians
2. Aves *(Birds)*
3. Mammals
4. Reptiles
5. Pisces *(Fish)*

Fish are actually divided into four classes, making eight classes of Chordata. Pisces once referred to all fish, but is now generally used to refer to bony fishes (Osteichthyes). The four classes are taught at the high school level (Agnatha, Placeodermi, Chondrichthyes, Osteichthyes), but would be needlessly confusing at the intermediate and junior high levels.

Continue to look at specific examples within each category. Include:

1. *Classification*

2. *Structure* (diagram or description)

3. *Growth and development*: birth to death (life cycle), reproduction, life span.

4. *What it needs for life*: environment, diet

5. *What eats it*, if anything. (predators)

6. *Places* where it is most common. (habitat and locations geographically)

 *Protozoa* are commonly studied while examining ecosystems and examining pond water under a microscope. (Many live as parasites in fresh and salt water and soil.) Diagrams of the protozoans, amoeba or paramecium may be examined during studies of the cell. Students should understand how water and nutrients are taken in and wastes eliminated, as well as reproduction by mitosis (cell division).

 *Echinoderms* live only in salt water. Therefore, examples are often introduced as part of an ocean habitat. Starfish are popular because they can lose a limb and grow a new one (regeneration).

 Porifera are identified by holes (pores) all over their bodies and inhabit warm ocean waters. The natural sponges used for bathing are the skeletons left behind. They could easily be studied as part of an ocean habit, or out of curiosity when using a sponge for cleaning. (*Be sure young children realize the difference between natural and synthetic sponges.*)

HIGH SCHOOL

Continue to study the Biological Classification System with the understanding that it is based on similarities and differences in structure, and on opinion.

Examples and explanations now include: **Kingdom, Phylum, Class, Order, Family, Genus, Species**

Vertebrates are now listed under **Kingdom**: *Animalia,* **Phylum**: *Chordata* and **Subphylum**: *Vertebrata.* Continue to study specific examples, finding answers to the questions listed on the previous page (Intermediate - Junior High section), including detailed explanations of respiration and reproduction. Vocabulary becomes more precise at this level. Dissection may aid a student in studying internal systems of an animal.

Most vertebrates have an endoskeleton made up of bone and cartilage. However, some fish have skeletons made up of cartilage only.

Mammals: Include a definition and examples of marsupials and explain the difference between a marsupial and placental (young nourished inside the body in a placenta). Define the order Primate and give examples.

Porifera: Students should be aware that there are thousands of species. In looking at a specific example, questions are answered in terms of groups of cells carrying out a specific job—involved in the process for gathering food, providing structural support, and so on. Vocabulary: *pores, spicules, collar cells, regeneration.*

Coelenterates: Identify the two different body types—medusa and polyp. Examine the need for stinging cells (nematocysts) on the tentacles and other body parts, and realize that only coelenterates have them.

Classes of Fish: Agnatha, Chondrichthyes, Osteichthyes and Placodermi (extinct armored-fish with jaws). Some texts only include three classes of fish since those that make up the class Placodermi are extinct.

Agnatha	This class includes fish without paired fins or jaws, such as lampreys. *(Lampreys are snake-like in looks and suck body fluids from victims.)*
Chondrichthyes	Sharks and stingrays have a skeleton made up of cartilage and belong to this class
Osteichthyes	These are the bony fishes (the largest class) and include most of the fish we eat or observe in home aquariums.

VERTEBRATES
backbone, cranium, separate bones, organs
phylum: **Chordata**

AMPHIBIANS   cold-blooded, mucous covering, lay eggs, gills before lungs

BIRDS   warm-blooded, covered with feathers, eggs have hard shells, 4-chambered heart

FISH   cold-blooded, gills, covered with scales, have fins in pairs, eggs are fertilized after laying, 2-chambered heart

MAMMALS   warm-blooded, lungs (breathe air), some hair, live babies, make milk and suckle their young

REPTILES   cold-blooded, covered with dry scales, short or no legs, lungs (no gills), 3-chambered heart

INVERTEBRATES
no backbone, cold-blooded, small, soft bodies
classified by phylum

ARTHROPODS   jointed legs; outer, armor-like skeleton

Classes:

Arachnids	*spiders, scorpions, horseshoe crab*	**Centipedes**
Crustaceans	*shrimp, lobster, barnacles*	**Millipedes**
Insects	*bees, grasshoppers, termites, ants*	

PORIFERA   pore-bearing sponges

COELENTERATE   hollow cavity used for digestion and circulation
corals, jellyfish, hydra, sea anemones

ECHINODERMS   spiny-skins, radial symmetry
starfish, sea urchin, sand dollar

MOLLUSKS   soft-bodied, limey shells
snail, clam, oyster, octopus, squid, scallop

WORMS

flatworms	(phylum: platyhelminthes)
round worms	(phylum: nemathelminthes)
segmented worms	(phylum: annelida)

PROTOZOAN   single - cell *(may be classified as Protist)*

ANIMALS: Life Cycle

Animals reproduce their own kind.

PRIMARY

▶ Match animal babies to their parents.

▶ Observe that some babies hatch from eggs, others are born live.

INTERMEDIATE - HIGH SCHOOL

▶ Study the life cycle (stages of development) of at least one animal in each category, observing that the babies always grow into adults with the same characteristics as the parents. *(The butterfly, salmon, and frog are often studied.)*

▶ Observe babies of parents that are within the same classification, but with enough differences to observe various results of genetics. *(mule)*

JUNIOR HIGH - HIGH SCHOOL

▶ Identify parts of an animal cell.

▶ Explain the function of each part of an animal cell:
- ◆ Cell membrane
- ◆ Cytoplasm
- ◆ Endoplasmic reticulum
- ◆ Mitochondrion
- ◆ Nucleus
- ◆ Nuclear membrane
- ◆ Ribosomes
- ◆ Vacuole

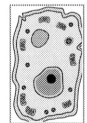

▶ Explain the role of DNA and RNA chromosomes and where to find them.

▶ Identify a diagram of a DNA molecule, pairing bases to form "rungs" on the DNA "ladder."

▶ Identify the stages of mitosis *(cell reproduction)*.

⇒ Explain what occurs at each stage: prophase, metaphase, anaphase, telophase

HIGH SCHOOL

Look at reproduction beginning with fertilization—follow the entire development.

ANIMALS: Habitats

Animals need certain things in their environments in order to live.

PRIMARY - INTERMEDIATE

Observe, explore, and discover, how different animals:

▸ **Get food**. What food do they need?
(Children should discover that each animal needs particular foods, and that all animals do not eat the same things.)

▸ **Protect themselves**.
(Children should observe the kind of shelter an animal makes or finds in order to raise its young.)

⇒ Ask whether or not the way the animal moves helps it protect itself.

⇒ Look for adaptations to the surroundings. (color, shape)
(praying mantis, chameleon, butterfly)

⇒ Look for ways animals defend themselves.
(odor of skunk, claws, quills, poison, hide in shell)

⇒ **Adapt to changes in season.**
(change of color, thickness of coat, migration, hibernation)

▸ While discovering how animals adapt to their surroundings, observe the type of habitat that makes up its environment: water (fresh or salt), forest, desert, tropics, grasslands, or cold lands. Also look at its predators. This "sets the stage" for learning **how the following adaptations are helpful:**

♦ **Living in a group** (if applicable)

♦ **Body parts**
⇒ Type of *teeth* (sharp and pointed, wide and flat) or type of *bill* in relation to *what it eats and how.*
⇒ Types of *feet* and why that type is helpful. (Also, legs in the case of birds, fins only in the case of fish.)

- ◆ **Body shape**

- ◆ **Body covering**
  - ⇒ What is down? Why is it helpful? Why feathers instead of hair?
  - ⇒ Why different colors in different seasons? and so on.

- ◆ **Defense mechanisms**
  - ⇒ Odor released by skunk, quills of a porcupine, poison, camouflage

- ◆ **Cold-blooded or warm-blooded**

- ◆ **How they breathe**
  - ⇒ Respiratory systems: gills/lungs/air tubes

- ◆ **Migration**

- ◆ **Hibernation**

JUNIOR HIGH - HIGH SCHOOL

Continue with the material listed above and add **Patterns of Animal Behavior**:

- ▶ Inborn behavior (periodicity - daily cycles in activities, territoriality, communication)

- ▶ Learned behavior

ANIMALS: Ecosystems / Biomes

Animals and plants depend on each other.

PRIMARY

- ▶ Animals eat plants, or eat other animals that eat plants.

- ▶ Animals use plants for shelter.

- ▶ Plants need the wastes of animals and the bodies of dead animals to replenish the soil.

INTERMEDIATE - JUNIOR HIGH

- ▶ Explain the role of predators and scavengers and name examples.

▸ Explain what is meant by the "**Balance of Nature**."

▸ Define ecosystem.

▸ Study animals in their environments.

Activities

1. Keep an ant colony for observation .

2. Observe fish in an aquarium.

3. Roll away a decaying log to discover (observe) the types of life that live there.

4. Observe birds in the park, or build a feeder and watch birds in the yard.

5. Conduct experiments while studying common insects. Test earthworms to find out if they prefer the light or the dark, warmth or cold, etc.

JUNIOR HIGH - HIGH SCHOOL

▸ Define and understand the differences between the following:
   ♦ Community -   Population
   ♦ Ecology   -   Ecosystem
   ♦ Food chain -   Food Web
   ♦ Habitat   -   Niche
   ♦ Predator   -   Prey

▸ Habitats (listed under plants: desert, ocean, forest, pond) are redefined:

BIOMES are regions characterized by a specific climate and plant and animal life.

Junior High:	**Land Biomes**	Coniferous Forest
		Deciduous Forest
		Desert
		Grassland
		Rain Forest
		Tundra
	Water Biomes	Freshwater
		Marine

High School: **Land Biomes**

Mid-latitude deciduous forest
Mid-latitude desert
Mid-latitude grassland
Mid-latitude rain forest
Taiga,
Tropical deciduous forest
Chaparral (term used in America),
and mountains as exceptions
Tropical rain forest
Tundra

▶ **A look a biomes includes:**
  ♦ Climate (*wet, dry, cold, hot, growing season, etc.*)
  ♦ General vegetation.
  ♦ Typical wildlife
  ♦ Latitude and longitude   (Find the region on a map.)
  ♦ Understanding of "succession."
    That is, learning how ecosystems move toward a "climax" or the one ecosystem that is determined by the climate. For instance, a forest fire destroys the "climax" ecosystem (the forest), but not forever. Small plants begin to grow, life continues, gradually achieving that climax ecosystem once again.

▶ **Terms to define:**
  ♦ Primary succession
  ♦ Secondary succession
  ♦ Pioneers
  ♦ Climax community

▶ Students should also become familiar with factors that may cause succession and explain the affects on the food chain.

Natural disasters   Fire
                          Flood
                          Earthquake
                          Plant diseases

Interference by man   Clearing land and/or filling in ponds to build
                          Pollution from a variety of sources including pesticides on crops
                          that eventually make their way into water supplies

▶ Within biomes, individual ecosystems may be studied. *That is, the relationship of the living and nonliving things interacting in a specific area. That area may be as small as under a log in the forest, or a stream in someone's backyard.*

Ecosystems in water are divided into:
  ♦ Ponds
  ♦ Lakes
  ♦ Flowing Waters
  ♦ Oceans

▶ **While studying any of the ecosystems:**

⇒ Diagram a food chain using the following labels next to appropriate plants and animals:
- ◆ Producers
- ◆ First-level consumers
- ◆ Second-level consumers
- ◆ Third-level consumers
- ◆ Fourth-level consumers

⇒ Define
- ◆ Carnivore
- ◆ Herbivore
- ◆ Omnivore

⇒ Diagram a food web.

⇒ Diagram a food cycle including: producers, consumers, decomposers.

▶ **Besides the predator-prey relationship, students should recognize other relationships:**

Symbionts - Different kinds of organisms living together.

Types of symbiotic relationships:

Commensalism	One organism benefits from another organism which doesn't benefit, but isn't harmed. *(Insect eating food another insect has stored)*
Mutualism	Both organisms benefit from each other. *(Fungus and algae are plants so dependent on each other that together they are called lichen.)*
Parasitism	One organism benefits, the other does not, and may be harmed, but usually is not killed by the parasite.
Saprophytic	An organism uses the remains of a dead organism as food. *(Bacteria involved in decomposition—decay.)*

▶ Students should be familiar with the famous experiment of Francesco Redi in which he disproved the common belief (17th century) that rotting meat turned into maggots.

CYCLES

The following cycles may be studied as part of earth science, physical science, or in relation to biomes or ecosystems (life science):

Water Cycle

Water evaporates (because of the sun's energy) from ponds, lakes, rivers, oceans, and soil, That water, in the air, is cooled (condensation). At some point water in the clouds comes back to the earth as precipitation. The water seeps into the ground, adds more to lakes, seas, and such, and is taken in by plants and animals. Plants give up water through their leaves, animals through their wastes. Water evaporates once again, and the cycle continues.

Carbon Dioxide - Oxygen Cycle

Plants take in carbon dioxide, using it to make food, and release oxygen. Animals take in oxygen, and give off carbon dioxide. Decaying organisms and burning fuels also give off carbon dioxide. This carbon dioxide is taken in by plants and the cycle continues.

Nitrogen Cycle

Nitrogen in the air that is combined with other elements, turning it into a compound, is used by plants to make proteins. (Lightning causes nitrogen in the air to combine with either hydrogen or oxygen.) When it rains, these compounds, ammonia and nitrates, enter the soil to be used by plants and bacteria. Animals eat plants, taking in "fixed" nitrogen (nitrogen in a compound). Animals release it back to the soil through their wastes or when their bodies die and decomposers become active. Bacteria (called denitrifying bacteria) in the soil change the nitrogen compounds into nitrogen gas, which is then released into the air. That gas is combined with oxygen or hydrogen, and the cycle continues.

A look at ecosystems and habitats can also lead naturally to a study of resources and the preservation of the environment. (See "Earth Science" for specifics.)

PLANTS

Categories: Living and Nonliving

Plants are living things.

PRIMARY

Discover characteristics that separate plants from nonliving things.
(Growth may be the only idea offered at this age.)

INTERMEDIATE - HIGH SCHOOL

See *ANIMALS: Categories - Living and Nonliving*

JUNIOR HIGH - HIGH SCOOL

▶ Identify and explain the function of each part of a plant cell:
   ◆ Cell membrane
   ◆ Cell wall
   ◆ Chloroplast
   ◆ Cytoplasm
   ◆ Nuclear membrane
   ◆ Nucleus
   ◆ Vacuole

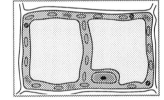

▶ Define
   ◆ Chlorophyll
   ◆ Organelle
   ◆ Plastid

▶ Describe similarities and differences between a plant and an animal cell.
   (Only plant cells have a cell wall and chloroplast.)

▶ Compare the processes of respiration and photosynthesis.
   (Photosynthesis uses energy, respiration releases energy.)

PLANTS: Distinguishing Characteristics

There are many kinds of plants, but with characteristics that allow us to tell them apart.

PRIMARY

▶ Observe similarities and differences between plants and animals.

▶ Observe similarities among plants easily observed by the student.
   (roots, stems, leaves)

Note: Children may not always recognize a stem. Ask "What is the stem's job?" *(Support, connecting the roots and leaves.)* Then, point out a tree. Ask, "What part does the same job?" This will lead to the understanding that the trunk is the stem of a tree.

All plants don't flower and produce seed at the same time. Children may draw the conclusion that those that don't have flowers when they observe them, never flower. You could ask whether or not the plant ever has flowers to help them realize that their observations are not complete.

▶ Observe differences among plants.
   ◆ Size
   ◆ Shape of leaf
   ◆ Different flower
   ◆ Different root
   ◆ Different shape or coating of seed
   ◆ Color
   ◆ Height
   ◆ Edible or not
   ◆ Grow from bulb/seed

▶ Identify common edible plants and seeds. *(fruits, vegetables, and nuts)*

▶ Observe seeds in plants we eat.

▶ Classify common foods as fruits or vegetables.

Note: Some common foods referred to as vegetables have seeds inside and are fruits.

▶ Name common trees and flowers.
   Have children observe that they can use the leaf shape as a key to naming the tree.
   Have them identify a variety of trees this way.

INTERMEDIATE - JUNIOR HIGH

At this level, a popular classification system is: **Flowering and Nonflowering plants**.

Classification by structure is based on the ideas of a Swedish botanist, Carl von Linne. He also divided all living things into two kingdoms: Plants and Animals. Classifying living things is called taxonomy.

 A look at Linne (or Carolus Linnaeus, his Latin name) should include his system used for naming living things: a two part name using Latin or Greek words that identify its place. In the animal kingdom, the first name tells the genus and the second tells the species.

Flowering Plants (class: *angiosperms*): All green plants that bear flowers.

Within this classification are two subdivisions, based on the number of leaves found in the seed of the flowering plant.
(Food is supplied by these leaves until the plant can make it on its own.)

1. **MONOCOTS**   (*Monocotyledons*):   one seed leaf

2. **DICOTS**   (*Dicotyledons*):   two seed leaves

Nonflowering Plants Plants that never flower.

1. **CONIFERS** (cone-bearing): Seeds are in cones. The leaves are usually needle-like.

   (This is an order of the class *gymnosperms*: seeds are attached to a scale.)

2. **NO SEEDS**: Do not reproduce by seeds. (spores, cell-division) *ferns, algae*

   ⇒ Although no longer classified as plants, a study should also include fungi, lichen, and molds.

JUNIOR HIGH - HIGH SCHOOL

A classification system used after the one by Linne (or Linneaus), which is also by structure, but not just reproduction:

PHLYA:

1. **Tracheophytes**: vascular system (tubes) extending throughout its roots, stems, and leaves
   (most plants)
   ▸ Some reproduce by seeds: **spermatophytes** (angiosperms and gymnosperms)

   ▸ Some reproduce by spores: **pteridophytes** (ferns)

2. **Bryophytes**: multicellular, simple leaves, parts that look like stems and roots but are not vascular, contain chlorophyll, make their own food by photosynthesis (*mosses, liverworts, hornworts*)

PLANTS: Habitats

Plants adapt to their environment.

PRIMARY

▸ Match a plant to its habitat: ocean, desert, forest, or pond.

▸ Find out how a few common plants react to changes in seasons.

▸ Sort leaves:

⇒ **Simple**   one leaf on the stem.
   Compound   more than one leaf on one stem.

⇒ Leaves with edges that are
♦ smooth
♦ serrated (like the jagged edge of a saw blade)
♦ lobed   (like an oak leaf)

INTERMEDIATE

▸ Identify plants as belonging to one of the major habitats: desert, forest, pond, ocean, (or fresh water and salt water instead of pond and ocean).

▸ Identify ways plants are suited to their specific habitat:
♦ Leaf size and shape
♦ Protective covering
♦ Root system
♦ Soil conditions

▸ Classify leaves by **shape** (*simple* or *compound*), **edges** (*entire* or *smooth, serrated, lobed*) and **veins** (*parallel* or *palmate*).

▸ Classify flowers and leaves as monocot or dicot.

Monocot: flowers have 3 petals, or multiples of 3; leaves have parallel veins

---

Dicot: flowers have 4 or 5 petals, or multiples of 4 or 5; leaves have branching veins

▸ Identify adaptations that may protect the plant from being eaten by animals. *(thorns, poisonous, bad taste)*

▸ Study the life cycle of some plants in each major habitat.

▸ Classify common trees as deciduous or evergreen.

JUNIOR HIGH - HIGH SCHOOL

Students should become acquainted with the problems in classifying plants, and the changes that have been made since the botanist Carl von Linne proposed the system based on reproduction in the 18th century. A recent system divides living things into five kingdoms: Animal, Plant, Protist, Fungi, Monera (See "Animals.")

Classification headings for plants are:

Kingdom   **Phylum**   **Subphylum**   **Class**   **Subclass**

PLANTS: Uses

Plants and animals depend on each other.

PRIMARY

▸ Discuss uses for plants: food, shelter, to build a fire.

▸ Diagram a FOOD CHAIN when discussing plants as food.

Lead children to the discovery that when we eat animals, we are still depending on plants because the animals eat plants.

INTERMEDIATE - JUNIOR HIGH

▸ Discuss uses of plants:
  ♦ Food
  ♦ Shelter
  ♦ To build a fire
  ♦ For pleasure
  ♦ To provide oxygen

▸ Explain the life cycle of a tree through to use.

⟹ Cut the trees → Transport the logs → Cut the logs → Sell the lumber

⟹ Look at uses for the lumber, including the process for making paper.

▸ Explain life cycle in general. *(diagram)*

Seed → *(grows)* Plant → Death → Decay *(decomposition)* → Enriching soil *(humus)* → A seed from the plant begins to grow, continuing the cycle.

▸ Explain plants as food.
(The food green plants make and store provides people and animals with nutrients.)

▸ Describe how plants make nutrients.
(Plants make sugars, usually glucose, and can change sugars into starches, fat, and proteins.)

◆ The process by which green plants make food is called **photosynthesis**.

yields
carbon dioxide + water + light *(energy)* → glucose + oxygen

◆ Plants are made of compounds that began as sugar. *(bark, petals, pollen)*

◆ These sugars and starches are carbohydrates: compounds of carbon, hydrogen, and oxygen.

◆ Plants may store this food in various places: roots, stems, leaves, or fruit.
We eat the part of the plant in which the food is stored.

▸ Conduct experiments to identify sugar and starch in plants we use as food.

▸ Sort foods according to where the food is stored (which part we eat):
root - carrot   **leaves** - spinach, lettuce
stem - celery, asparagus   **fruit** - apples, peaches

 Potatoes are stems— the "eyes" are its buds. Sweet potatoes and yams are roots.

▸ Diagram the carbon dioxide - oxygen cycle.
(Study this while learning about the process of photosynthesis.)

Plants take in carbon dioxide for photosynthesis and give off oxygen. The carbon dioxide comes from animals and humans breathing (out), plants and animals decaying, and plants burning. (Animals take in the oxygen.)

▶ Explain the role of plants in the **FOOD CHAIN**.

 Only plants can make their own food (photosynthesis) and are therefore the first link in a chain.
(Look at algae or plankton as the first in one food chain example.)

▶ Illustrate food chains using a variety of plants and animals, and representing various plant habitats.

PLANTS: Growth

Plants need certain things in order to grow.

PRIMARY

▶ Identify a plant's needs.
  ◆ A plant begins as a seed and needs air, water, soil and warmth to grow.
  ◆ Different plants require different environments.
  ◆ Plants differ in the amount of light, water, and type of soil needed.

▶ Explain how a plant grows. (cycle: seed to adulthood to decay)

▶ Determine the job of each part of a plant:
  ◆ Flower or cone
  ◆ Fruit
  ◆ Leaf
  ◆ Root
  ◆ Seed
  ◆ Stem

Activities

1. Plant seeds. (Commercial potting soil gives the best results for germinating seeds.) Put plants in various conditions to discover what happens if there is too little or too much sun, air, water, and so on.

2. Put a stalk of celery (one with leaves) in a clear glass with water that has food coloring added. Over a few days, children can observe the dye moving up the celery stalk and eventually into the leaves. This allows them to see how water travels from roots to leaves in a vascular system.

▶ Explain requirements for plant growth:
  ◆ Minerals from the soil. (*nitrates, phosphates*)
  ◆ Water to dissolve the minerals.
  ◆ Oxygen and carbon dioxide from the air. (*Some don't need oxygen.*)
  ◆ Light for photosynthesis.

(Here the earth science of how soil is formed and what it contains could be included, as well as a look at the nitrogen cycle.)

 While examining plants in various habitats, include the differences in soil. Help students understand that plants differ as to what they need from the soil.

▶ Explain the process of photosynthesis (how plants make their own food). (Older students should be able to write it as an equation.)

▶ Define and explain the role of chlorophyll.

▶ Identify plant parts and explain their functions:

Leaf   upper and lower epidermis, palisade layer, vein, guard cells, chloroplasts, stomata, sheath, vascular cells (xylem and phloem)

Stem   Plant stems vary but most have leaves and fibrovascular bundles (groups of vein tubes).

  ◆ Classification of stems: monocot, dicot
  ◆ Layers of a woody stem: bark, cambium, wood, pith

Root   xylem, phloem, cortex, epidermis, root hairs, taproot, secondary roots (brace roots)

  ◆ Function: store food, absorb water and minerals from soil, hold the plant in place

(A few texts include "root cap," which is at the end of the root and protects the tip as it pushes through the soil.)

Flower parts   petal, pistil (stigma, style, ovary), sepal

Cell   cell wall, chloroplast, (which are not found in animal cells), nucleus, chromatin (chromosomes), cytoplasm, golgi bodies, vacuoles, mitochondria

Xylem cells   carry water and minerals

Phloem cells   carry food

▶ Describe the function of each part of a plant cell.

▶ Diagram a cell. Label each part.

▶ Describe the difference between a plant and animal cell.

JUNIOR HIGH - HIGH SCHOOL

▶ Explain how plants grow (cell division).

▶ Identify each phase of mitosis, explaining what occurs at each stage.

▶ Identify a model of a DNA molecule. Match base pairs.

PLANTS: Reproduction

Plants reproduce their own kind.

PRIMARY

▶ Observe and study plants which reproduce with seeds.

| Activities | **Learn about edible seeds:**

1. Grind seeds into flour. (corn, wheat, rye, rice)

2. Toast seeds for a snack.
   (pumpkin, sunflower, squash)

3. Cook seeds for a meal or snack. *(rice, oatmeal, popcorn)*

4. Grind peanuts into peanut butter.

5. Open a coconut (one of the biggest seeds): drink the liquid, taste the "meat."

INTERMEDIATE - JUNIOR HIGH

▶ Classify plants as flowering or nonflowering.

► Classify common trees as gymnosperms (evergreens) or angiosperms (leafy trees).

⇒ How are they different?
♦ Flowers   -   No Flowers
♦ Needles   -   Leaves
♦ Seeds in cones - No cones   (*and so on*)

► Classify plants by method of reproduction: Seeds/No Seeds

 **Plants that reproduce by seeds:**

CONIFERS   (*seeds in cones*)

⇒ Name common trees that reproduce with seed in cones.
(*evergreens such as pine and spruce, a few deciduous trees*)

⇒ Describe the process of reproduction, including fertilization.

FLOWERING PLANTS   (*seeds inside fruit*)

⇒ Identify parts of a flower and explain the function of each part:
♦ Eggs
♦ Ovules
♦ Petals
♦ Pistils
♦ Pollen Grains
♦ Pollen Tube
♦ Stamen
♦ Sepals

⇒ Carefully dissect a flower, naming the parts.
Ovule   -   Egg Cell
Pistils   -   Female Organ
Pollen grain -   Sperm cell
Stamen   -   Male organ

⇒ Explain how seeds are made.

► Describe the process of reproduction in:

♦ Self-pollination (*carrying the pollen from the anther to the stigma in the same flower*)

◆ Cross-pollination *(carrying the pollen of the anther of one flower to the stigma of another plant)*

 Students could begin by hypothesizing how the pollen gets to the pistil and then study the interdependence of flowers and bees.

▶ Define:

fertilization	uniting sperm cell with egg cell
seed	embryo and stored food, covered by a seed coat
germinate	grow or develop
dormant	stopped developing
cotyledons	leaf or leaves of seed which contains stored food
fruit	enlarged, developed ovary

 At the junior and senior high levels look at foods to determine if they are fruits or vegetables. Many foods commonly referred to as vegetables are actually fruits. Students should be able to explain why a food is classified as a fruit.

▶ Explain the development and purpose of fruit.

▶ Identify ways seeds are spread. *(wind, animals, burrs)*

 Plants that grow without seeds:

PLANTS THAT GROW FROM PARTS OF THE STEM OR ROOT

Grow a plant from a piece of a fresh carrot or potato.

PLANTS THAT REPRODUCE WITH SPORES: Ferns, Mosses, Algae, Fungi

FERNS

◆ Are vascular. *(They have stems with a system of tubes inside.)*
◆ Have roots and leaves. *(fronds)*

MOSSES

◆ Small, leaf-like structures grow on a central stalk.
◆ Root-like structures take up water and minerals from the soil.
◆ Reproduction can be accomplished by forming spores and by forming sperm cells and egg cells.

Their usefulness may be studied along with earth science. They help form soil, are good ground cover and absorb large amounts of water (preventing soil from washing away). They often grow on rock surfaces.

ALGAE *May be listed as a Protist, not as a plant.*
- ♦ Contain chlorophyll (but not all types are green).
- ♦ Carry on photosynthesis.

 One type, plankton, is often studied because of its importance in the food chain.

⇒ Place algae in a food chain.

⇒ Seaweed and kelp are algae: identify their needs.

 Notice that they grow in shallow areas in order to receive the light from the sun that is necessary for photosynthesis.

FUNGI *Now listed as a Kingdom of its own, rather than part of the Plant Kingdom.*
- ♦ Do not have chlorophyll and cannot make their own food.
- ♦ Live off of dead or decaying things (are either saprophytes or parasites).
- ♦ Are nonvascular (no tubes to carry food and water).
- ♦ Reproduce with spores, not seeds.

Common examples of fungi: mushrooms, mold yeast, rust mildew

⇒ Identify characteristics of mushrooms, molds, and yeast, and the environment each needs for growth. Include the process of reproduction of each.

⇒ Define mycelium (on mushrooms) and their function. (to find food)

⇒ Explain yeast budding.

⇒ Explain how yeast helps dough rise (how it gives off carbon dioxide)

⇒ Observe and experiment with molds and yeast.

⇒ Explain the use and harm (if any) of mushrooms, molds, and yeast. Include:
- ♦ Making cheese (*mold*)
- ♦ Bread and alcohol (*yeast*)
- ♦ Milk products such as yogurt (*bacteria causing milk to sour*)
  Bacteria was at one time included with fungi as a protist
- ♦ Decomposition to enrich soil

⇒ Describe the discovery of penicillin and explain what it is and what it does.

 Lichen may be observed while studying habitats, or weathering. It is a combination of algae and fungus living together and helping each other (symbiosis). It can grow in the Tundra where almost nothing else can. It grows on rock, slowly breaking the rock down into soil.

HIGH SCHOOL
Continue studying the above in addition to the following:

▶ Determine phylum and class. (Classify)

▶ Experiment with protists. Make cultures.

▶ Study bacteria (Monera Kingdom) and viruses. Include:

⇒ The process of reproduction and growth.

⇒ How a virus spreads.

⇒ Useful and harmful bacteria. (Some texts introduce this at the junior high level.)

⇒ A definition of microbes and microbiology.

⇒ Contributions of Louis Pasteur and Robert Koch.

⇒ The process of pasteurization of milk.

⇒ The classification of diseases and a study of how they spread.

⇒ Definitions of the following:
  ◆ Antibodies
  ◆ Contagious
  ◆ Epidemic
  ◆ Host
  ◆ Immunity
  ◆ Parasite
  ◆ Pathogen
  ◆ Resistance
  ◆ Spontaneous Generation (including background of Anton van Leeuwenhoek)
  ◆ Virulence

JUNIOR HIGH - HIGH SCHOOL

▸ Identify the following terms:

1. **Circadian rhythm**   Whatever would be considered the plant's normal activity during a twenty-four hour cycle.

2. **Negative geotropism**   The plant's stem grows upward, away from the direction of gravity

3. **Phototropism**   The plant's growth toward the source of light.

4. **Positive geotropism**   The plant root grows downward—the direction of gravity.

5. **Touch response**   A plant's movement in response to being touched.

At earlier levels, experiments with plants may lead to observations of any of the above, but these terms may not be used for what has been observed until later.

THE HUMAN BODY

ANATOMY: Parts of the Body

PRIMARY

▸ Name the following body parts:
- head   neck   face   eyes   eyebrows   eyelashes   nose
- mouth   lips   teeth   tongue   chin   cheeks   hair
- shoulder   arm   elbow   hand   wrist   fingers   fingernail
- chest   waist   hips   leg   thigh   knee   calf
- ankle   foot   toes   toenails   skin   (Foot: ball, arch, heel - grade 3+)

▸ Observe physical differences among people: hair color and texture, eye color, skin tone

▶ Notice body symmetry:
If you draw a vertical line down the center, each side looks the same.

Activity

Trace the child's body outline on butcher block paper, then have him draw in his face, hair, fingers, and clothing.

ANATOMY: The Five Senses

sight, hearing, taste, touch, smell

PRIMARY

▶ Identify the five senses.

▶ Identify the interrelationship of taste and smell.
(Have children taste samples blindfolded, then blindfolded *and* holding their noses to prevent smelling what they taste.)

▶ Identify taste buds and the parts of the tongue that respond to sweet, bitter, salty, and sour.

INTERMEDIATE

▶ Observe that if the tongue is wiped dry, salt or sugar can't be tasted: they must be dissolved first.

▶ Observe that tongues are wet with saliva (which dissolves food).

▶ Identify the role of the skin and nerve endings for a sense of touch.
(*hot, cold, pressure, pain, texture*)

▶ Observe which areas of the body are the most sensitive to touch and find out why.

JUNIOR HIGH - HIGH SCHOOL

▶ Explore touch more thoroughly during study of the nervous system:
sensitivity to pressure, temperature and pain.

ANATOMY: Ears and Sound

PRIMARY

▶ Observe that we hear sounds through our ears.

▶ Name parts of the ear and tell what they do:
- ◆ Inner ear
- ◆ Middle ear
- ◆ Outer ear
- ◆ Nerve to brain

▶ Identify sounds.
- ⟹ Identify a sound as loud or soft.

- ⟹ Identify what made the sound. (bell, lawn mower, musical instrument, hands clapping)

- ⟹ Identify the location of the sound (direction).

- ⟹ Identify the pitch by describing the sound as high or low.

▶ Observe that sounds are made by vibrations.

- ⟹ Sounds are vibrations which travel through many materials.

- ⟹ Ears help us hear sounds.

▶ Observe that sounds can be heard more clearly under some conditions.

 Conduct experiments that allow sound a path to travel: 2 paper cups attached with a string; the tube from wrapping paper, and so on.

INTERMEDIATE - JUNIOR HIGH

▶ Identify causes of differences in pitch. (*length, thickness, tension*)

▶ Identify how sound travels:
- ◆ Moves through a medium, such as air (*So sound can't travel through outer space*)
- ◆ Moves through solids, liquids and gases
  (*Experiment to identify good conductors of sound.*)
- ◆ Is reflected (echo) or absorbed

► Explain movement in terms of: vibration   molecules   movement of energy   waves

► Explain how radio waves travel and how they are used.

► Define: sonar   ultrasonic sound   supersonic sound.

► Explain how bats use sound.

► Explain how dolphins use sound.

► Define acoustics and give examples of materials used.

► Define amplitude, identify amplitude on a graph and explain how to measure amplitude.

► Explain how to measure the speed of sound by measuring wavelengths.
Wavelength equals Speed divided by Frequency

► Make calculations of distance based on the speed of sound through water and air.

► Identify parts of the ear and what they do:
  ◆ Anvil
  ◆ Auditory Canal
  ◆ Cochlea
  ◆ Eardrum
  ◆ Eustachian Tube
  ◆ Fluid
  ◆ Hammer
  ◆ Inner Ear
  ◆ Middle Ear
  ◆ Nerve to Brain
  ◆ Outer Ear
  ◆ Semicircular Canals
  ◆ Stirrup

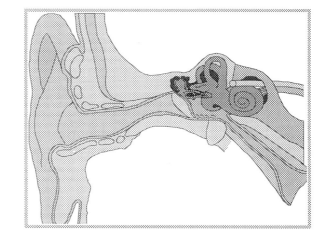

► Explain how the following work in relation to sound:
  ◆ Audio Tape
  ◆ Phonograph Record
  ◆ Telegraph
  ◆ Telephone

► Explain Michael Faraday's contribution to science.

ANATOMY: Eye, Sight, and Light

PRIMARY

▶ Identify parts of the eye and explain what they do: pupil   iris   lens   nerve

▶ Become familiar with how we see. (Diagrams should include the inverted image.)

INTERMEDIATE - JUNIOR HIGH

▶ Identify parts of the eye and explain what they do:
  ♦ Blind spot
  ♦ Ciliary muscle
  ♦ Cornea
  ♦ Convex lens
  ♦ Eyeball muscles
  ♦ Iris
  ♦ Liquid
  ♦ Optic nerve
  ♦ Pupil
  ♦ Retina (rods and cones)

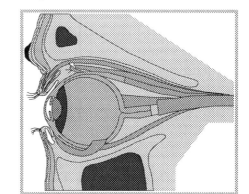

  ⇒ Explain the change in the pupil in response to light.

▶ Explain how we see, including the need for two eyes. (*depth perception*)
  Explanations should include inverted image and reflected light.

Activities helpful for understanding sight:
1. Compare the eye to a camera and make a pinhole camera.
2. Dissect an eyeball.
3. Make a periscope.

LIGHT

▶ Conduct experiments using: magnifying glass   mirror   prism

▶ Identify properties of light:
  ♦ Travels in straight paths.
  ♦ Can be absorbed.
  ♦ Can be reflected.
  ♦ Can be scattered.
  ♦ Can be bent.

- Define:
  - Opaque
  - Translucent
  - Transparent

- Identify ways light:
  - Acts like waves
  - Acts like particles
  - Is a form of energy
  - Can be measured as energy *(The unit is foot-candle.)*

 Energy is changed from one form to another to make light. The sun uses energy and gives off light, an electric bulb uses energy as it glows. Light that is absorbed is changed into heat energy.

- Identify ways to change the direction of light:
  Mirror
  Lenses: concave, convex, magnifying, lens of eye

- Define farsightedness and nearsightedness.

  ⇒ Illustrate how lenses can correct these.

- Define the spectrum of light.

  ⇒ Observe the spectrum by using a prism.
  (White light contains 7 colors: red, orange, yellow, green, blue, indigo, and violet)

- Explain how we see color. *(We see the reflected color.)*
  - Black: no color is reflected
  - White: all colors are reflected

- Define infrared and ultraviolet light.

- Calculate distance using the speed of light.
  186,000 miles per second through space
  140,000 miles per second through water

- Identify contributions to science by:
  - Christian Huygens
  - Anton Van Leewenhoek
  - Isaac Newton

ANATOMY: Teeth

PRIMARY

▸ Practice proper brushing and flossing.

▸ Explain the role of a dentist.

▸ Name two types of teeth. (*primary, permanent*)

▸ Identify a use for teeth. (*bite, chew*)

INTERMEDIATE - JUNIOR HIGH

▸ Identify parts of a tooth:
   ◆ Artery
   ◆ Cementum
   ◆ Crown
   ◆ Dentin
   ◆ Enamel
   ◆ Gum
   ◆ Nerve
   ◆ Pulp
   ◆ Periodontal Membrane
   ◆ Roots
   ◆ Veins

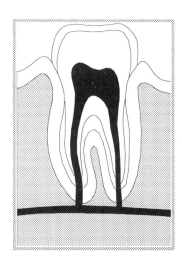

▸ Name the number of primary and permanent teeth.

▸ Name the different teeth and their function: (*bite, tear, grind*)
   ◆ Central and lateral incisors
   ◆ First and second bicuspids
   ◆ First, second, and third molars

▸ Define cavity and the causes. (*plaque, bacteria, acid, role of sugar*)

JUNIOR HIGH - HIGH SCHOOL

▸ Explain proper care of teeth including:
   Diet
   Brushing, flossing, fluoride
   Dental checkups, filling cavities, braces

▶ Define dental problems including:
  ◆ Dental abrasion
  ◆ Dental erosion
  ◆ Dentures
  ◆ Gingivitis
  ◆ Malocclusion
  ◆ Periodontal disease
  ◆ Periodontitis
  ◆ Trench Mouth

ANATOMY: Cell

INTERMEDIATE - HIGH SCHOOL

▶ Identify the cell as the basic unit or building block of living things:
It is alive and needs water, food, oxygen and a system to remove waste.

▶ Identify cell parts and explain their functions:
  ◆ Cell membrane
  ◆ Cilia
  ◆ Cytoplasm
  ◆ Nucleus

  Junior-Senior High:
  ◆ Centrosome (in animal cells)
  ◆ Chromosomes
  ◆ Endoplasmic reticulum
  ◆ Genes
  ◆ Lysosomes
  ◆ Mitochondria
  ◆ Nuclear Membrane
  ◆ Proteins
  ◆ Ribosomes
  ◆ Vacuoles

▶ Identify cell categories. (*covering cells, supporting cells*)

▶ Identify different types of cells. (*skin, muscle, bone*)

▶ Define tissue.

▸ Define organ.

JUNIOR HIGH - HIGH SCHOOL

▸ Identify five types of tissue in the human body and the function of each:
  ◆ Blood
  ◆ Connective
  ◆ Epithelial
  ◆ Muscle
  ◆ Nerves

▸ Explain the role of chromosomes.

▸ Describe how cells reproduce, identifying the stages of mitosis.

▸ Explain the replacement of cells, identifying those which are not replaced after a certain stage of development. (*brain, nerve*)

▸ Define and diagram a DNA molecule.

▸ Explain how scientists think DNA copies itself.
  (*DNA controls cell activities and should be studied along with cells and mitosis.*)

▸ Explain the role of RNA.

GENETICS / HEREDITY

This could be included in a thorough study of reproduction which includes a look at cells.

▸ Define:
  ◆ Blending
  ◆ Chromosome
  ◆ Dominant trait
  ◆ Fraternal twins
  ◆ Gene
  ◆ Genetic
  ◆ Genotype
  ◆ Heredity
  ◆ Hybrid
  ◆ Identical Twins
  ◆ Incomplete dominance
  ◆ Law of dominance
  ◆ Law of segregation
  ◆ Mutation
  ◆ Phenotype

- ◆ Physical Traits
- ◆ Punnett Square
- ◆ Purebred
- ◆ Recessive Trait
- ◆ Sex-Linked Trait

▶ A study of genetics should include information on Gregor Mendel and his experiments with plants. The terms to be defined (above) would be found in explanations of his experiments.

▶ Students should fill in Punnett squares to apply understanding.

 Theories of natural selection and evolution are often included in selections that provide information on genetics. Students should also study scientific evidence (written by scientists) opposing evolution.

ANATOMY: Skin

INTERMEDIATE - HIGH SCHOOL

▶ Define dermis and epidermis.

▶ Label a diagram that includes pores, oil glands, dermis and epidermis.

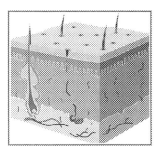

▶ Explain the use of pores and the need for perspiration.

▶ Describe the purpose of a scab.

▶ Identify causes and treatments of acne.

▶ Define suntan. (What is happening to the skin, and why?)

▶ Define first, second and third degree burns and describe treatment.

heart and blood vessels - transportation of materials throughout the body

INTERMEDIATE - HIGH SCHOOL

▸ Explain the job of the blood. *(carry food and oxygen to body parts, carry away wastes)*

▸ Identify the make-up of blood and the role of each:
Plasma   Red blood cells
Platelets   White blood cells

▸ Define monocyte and lymphocyte and explain their job.

▸ Define clotting and explain its importance. (Include platelets.)

▸ Describe the path the blood follows. Include the following:
◆ Arteries
◆ Capillaries
◆ Veins
◆ Body cells carrying food and oxygen
◆ Dark red and bright red blood

▸ Identify the parts and functions of the heart and their role in the circulatory system:
◆ Atria
◆ Lungs
◆ Valves
◆ Ventricles
◆ Vessels

▸ Explain how blood cells are made and replaced.
(Include bone marrow and spleen.)

▸ Define lymph, lymph vessels, and lymph nodes and then explain their function.

▸ Explain what a pulse is.

▸ Identify symptoms and causes of the following diseases of the blood:
◆ Anemia
◆ Edema
◆ Hemophilia
◆ Leukemia

ANATOMY: Digestive and Excretory Systems

INTERMEDIATE

 A study must include information that helps a student understand that food must be changed into molecules for it to be considered usable.

 Digestion begins in the mouth with the breaking down of food by chewing and mixing it with saliva. A study of teeth could be included here, as well as the role of the tongue.

▸ Define Alimentary Canal.

▸ Identify and explain the function of:
  ♦ Appendix
  ♦ Esophagus
  ♦ Gall bladder
  ♦ Large intestine (colon)
  ♦ Liver
  ♦ Mouth: teeth, tongue, saliva
  ♦ Pancreatic juices/how they get to the food
  ♦ Pancreas
  ♦ Rectum
  ♦ Small Intestine
  ♦ Stomach

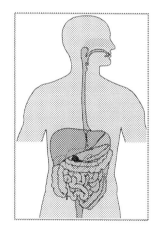

▸ Explain how the digested food reaches other parts of the body.
  (*villi, their location, role of the blood*)

▸ Explain how wastes are removed from the body. (*This is the excretory system.*)
  ♦ Bladder
  ♦ Intestines
  ♦ Kidneys
  Junior-Senior High:
  ♦ Ureters
  ♦ Urethra
  ♦ Urine
  ♦ Elimination of liquid wastes through breathing out (vapor contains water) and perspiring.

JUNIOR HIGH - HIGH SCHOOL

Along with the study of these systems, you may want to include the items listed under "Care of the Body" that apply to nutrition.

▶ Define and explain the role of:

Amino acids	Glycerol
Bile	Hydrochloric acid
Capillaries	Ileum
Digestive juices	Jejunum
Duodenum	Liver
Glucose	Lymph vessels
Enzymes	Mucus
Gall bladder	Pancreas
Gastric juices	Pancreatic juice

▶ Explain the body's need for sugar, starch, protein, and fat.

▶ Identify food sources for sugar, starch, protein, and fat.

▶ Food sources for necessary vitamins and minerals, and illnesses caused by vitamin or mineral deficiency.

ANATOMY: Endocrine System

regulates chemical activity

JUNIOR HIGH - HIGH SCHOOL

▶ Define gland. (This is a system of glands.)

▶ Give the definition, location, and function of:
   ◆ Adrenal glands
   ◆ Gonads: ovaries (female), testes (male)
   ◆ Pancreas
   ◆ Parathyroid gland
   ◆ Pituitary gland
   ◆ Thyroid gland
   ◆ Thymus

▶ These glands secrete hormones directly into the bloodstream which then travel throughout the body. Find out what each hormone does and what happens if too little or too much is secreted:

- ◆ Adrenal - adrenaline
- ◆ Ovaries - estrogen
- ◆ Pancreas - insulin
- ◆ Pituitary - growth hormone
- ◆ Testes - testosterone
- ◆ Thyroid gland - thyroxin

▶ Look at the role of the endocrine system in relationship to handling stress:

- ◆ Fight or flight stage (*role of adrenaline*)
- ◆ Resistance (*The body prepares to continue fighting or running, storing tensions in muscles if there is no fight or flight.*)

ANATOMY: Skeletal / Muscular Systems

support and movement

PRIMARY

▶ Identify a skeleton and describe its usefulness.
(A frame for the body may be the only concept understood at this level.)

▶ Classify muscles as voluntary or involuntary.

INTERMEDIATE - HIGH SCHOOL

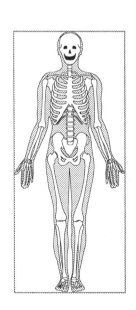

▶ Identify three functions of the skeleton.
(support, protect, allow movement)

▶ Describe movement as possible because of the interaction of the muscles and the skeletal bones.

▶ Identify main skeletal bones: (206 bones in the body)

Clavicle (collarbone)	Rib cage
Femur (thigh bone)	Scapula (shoulder blade)
Humerus (upper arm)	Skull
Patella (kneecap)	Sternum (breastbone)
Pelvis	Vertebral column (backbone)
Phalanges (finger and toe bones)	

▸ Identify which bones are protective (*skull, rib cage, vertebrae, hips*), which are used like levers to allow movement, and which are for support (*feet, hips, legs, back, neck*).

▸ Define bones:
 ◆ 3 layers (*marrow, cortex, outer layer*) and the purpose of each
 ◆ Shapes, sizes, how the size relates to the function
 Include understanding that bones are tissue made up of closely-packed cells.

▸ Define cartilage and explain its function:
 ◆ Heart muscle
 ◆ Skeletal (*for movement*)
 ◆ Smooth (*work inside the body*)

▸ Examine the role of the muscle organs:
 heart and lungs: size, shape, their parts and functions

▸ Examine the blood system in relation to the muscles.

▸ Suggest ways to build muscle endurance, strength, and flexibility.
 Explain why these are important. *(aerobic exercise, balance, coordination)*

▸ Develop a physical fitness routine that includes the following exercise principles:
 ◆ Warm-up
 ◆ Exercise just past tired (overload principle)
 ◆ Increase in intensity, speed, and time (progression principle)
 ◆ Cool-down exercises

Movement:
▸ Identify the following types of joints and demonstrate how each affects movement:
 ◆ Ball and socket
 ◆ Gliding
 ◆ Hinge
 ◆ Immovable joint

▸ Give an example of each joint on the body.

▸ Define ligaments. (*connecting bone to bone*)

▸ Define tendons. (*connecting muscle to bone*)

▸ Explain how muscles work in opposing pairs.
 (*one contracts, the other relaxes*)

 At the high school level, diagrams label the muscles of the body.

ANATOMY: Nervous System

receives and transports messages

INTERMEDIATE - HIGH SCHOOL

▶ Define and explain the function of:
- ◆ Nerves in relation to the five senses
- ◆ Central nervous system

▶ Identify nerve cells and their jobs:
- ◆ Connecting nerve cells
- ◆ Motor nerve cells
- ◆ Neuron
- ◆ Sensory nerve cells

▶ Diagram a neuron and describe the function of each part:
- ◆ Axon
- ◆ Dendrite
- ◆ End brush
- ◆ Nucleus

▶ Note that nerve cells cannot regenerate or repair themselves, like most body cells.

▶ Define: Reaction time
              Reflex action
              Voluntary and involuntary response

▶ Identify parts of the brain:
- ◆ Brain stem
- ◆ Cerebellum
- ◆ Cerebrum
- ◆ Medulla   (and the spinal cord)

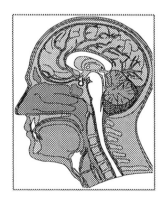

▶ Describe functions of the brain:
- ◆ Memory
- ◆ Moving with precision
- ◆ Reasoning (*and so on*)

▶ Describe disorders: Cerebral palsy
                          Paralysis

JUNIOR HIGH - HIGH SCHOOL

▶ Identify male characteristics:
Testes produce male sex hormones responsible for male characteristics.
(*beard and deep voice*)

▶ Identify female characteristics:
  ◆ Ovaries produce eggs
  ◆ Oviducts
  ◆ Uterus
  ◆ Menstruation

▶ Describe reproduction using the proper vocabulary:
  ◆ Egg fertilization
  ◆ Penis
  ◆ Semen
  ◆ Sperm
  ◆ Uterus
  ◆ Vagina

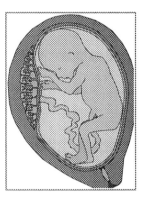

▶ Identify stages of development:
  ◆ Fertilization (embryo) through the nine months of growth.
  ◆ Nutrition of the baby within the uterus.

High School

▶ Define:
  ◆ Body cells
  ◆ Diploid cells
  ◆ Flagella
  ◆ Gametes
  ◆ Hermaphrodites
  ◆ Homologous chromosomes
  ◆ Meiosis
  ◆ Monoploid cells
  ◆ Sexual reproduction
  ◆ Zygote

▶ Identify the effects on the developing baby if the mother:
  ◆ Drinks alcohol
  ◆ Drinks coffee
  ◆ Smokes

▶ Explain the birth process and how to care for a newborn.

▶ Identify characteristics of the following stages of the life cycle:
  ◆ Infancy
  ◆ Childhood
  ◆ Adolescence
  ◆ Adulthood

▶ Study the stage of adolescence—define and explain causes and ways to cope with:
  ◆ Effects of nutrients (or lack of) on mental state
  ◆ Desire for independence
  ◆ Hormones
  ◆ Loneliness, anger, stress, need for identity
  ◆ Moods
  ◆ Puberty

▶ Compare and contrast ways of dealing with a problem:
  ◆ Facing the problem
  ◆ Focusing on and doing for others
  ◆ Keeping a positive mental attitude
  ◆ Relaxing
  ◆ Talking about it
  ◆ Vigorous exercise to relieve tension

  Include reactions such as:
    ⇒ Compensation
    ⇒ Daydreaming
    ⇒ Defense mechanisms
    ⇒ Denial
    ⇒ Displaced aggression
    ⇒ Projection
    ⇒ Rationalization
    ⇒ Repression
    ⇒ Suppression
    ⇒ Substitution

▶ Describe the effect of environment, diet, and exercise on growth and development.

ANATOMY: Respiratory System

breathing
(exchanges oxygen and carbon dioxide)

INTERMEDIATE - HIGH SCHOOL

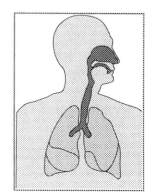

▶ Define, locate, and explain the function of:
- ◆ Nose
- ◆ Mouth
- ◆ Windpipe
- ◆ Lungs
- ◆ Diaphragm
- ◆ *High School:*
    - ◆ Alveoli
    - ◆ Bronchus
    - ◆ Capillaries
    - ◆ Trache

▶ Explain the role of the diaphragm muscle during breathing.

▶ Describe the respiratory process including oxygen (role of blood) and carbon dioxide (role of hemoglobin).

HEALTH: Care of the Body

Body types: ectomorph, mesomorph, endomorph

PRIMARY - INTERMEDIATE

Diet and Exercise

▶ Explain why we need exercise, rest, proper food (food groups), and clothing suited to the weather.

▶ Explain how exercise helps the body.
Blood supplies oxygen to the muscles.
The heart, which pumps the blood, is a muscle that grows stronger with exercise.

▶ Realize that food provides us with energy.

▶ Discover where our food comes from: growing or raising food, preparing and transporting food for sale, where food is sold.

▶ Classify common foods as nutritious or "junk" foods.

▶ Identify the following basic food groups:
  ◆ Bread-cereal
  ◆ Meat
  ◆ Milk
  ◆ Vegetable-fruit

▶ Become acquainted with our need for:
  ◆ A variety of foods
  ◆ A well-balanced diet
  ◆ Water

Activities

1. Field trips. (dairy farm, grocery store)
2. Cook recipes from other countries.
3. Sample new foods. (This may also involve studying the senses.)
4. Record foods eaten each day for a week to discover if your diet has been nutritious.

Grooming and Manners

▶ Practice proper daily care.

  ◆ Brush teeth
    ⇒ Recognize that toothpaste helps remove plaque
    ⇒ Use proper brushing technique
    ⇒ Floss permanent teeth

  ◆ Wash with soap: soap helps remove germs that we don't see

  ◆ Dress alone: choosing proper clothing

▶ Practice correct posture.
  (From the side—ears, shoulder, waist, knee, and ankles are in a straight line.)

  ⇒ Explain why it is important.
    It helps your blood circulate well, feeding oxygen to all parts. Poor posture crowds organs, makes circulation more difficult and may cause fatigue.

- ▸ Practice courtesy.
  - ◆ Polite behavior (words and actions)
  - ◆ Table manners
  - ◆ Respect for the property of others
  - ◆ Caring for property
  - ◆ Maintaining an orderly room. (bedroom, classroom)

Safety and First Aid

- ▸ Practice Safety rules:
  - ◆ Indoors
  - ◆ Outdoors   (Look both ways before crossing the street)
  - ◆ Using medicine
  - ◆ Using sharp objects
  - ◆ Using fire (matches)

- ▸ Practice simple first aid.

INTERMEDIATE - HIGH SCHOOL

- ▸ Describe the role of nutrition in maintaining a healthy body including:
  Food as fuel: calorie, basal metabolism, food oxidation

- ▸ Make-up of foods and the function of each:
  - ◆ The role of sugar and starch
  - ◆ Nutrient groups: protein, carbohydrates, fats, vitamins, minerals
  - ◆ The role of water

- ▸ Explain the specific function, and results of a lack of the following vitamins and minerals and cite examples of food sources for each:
  - ◆ Vitamins   A   C (absorbic acid)   D   E
                   B vitamins: thiamine, riboflavin, niacin, pantothentic acid, pyridoxine, cobalamin, folic acid
  - ◆ Minerals - calcium, iodine, iron, phosphorus, potassium, sodium

- ▸ Define amino acid and illustrate the molecular structure of one.

- ▸ Identify effects on the body of:
  - ◆ Alcohol
  - ◆ Candy
  - ◆ Narcotics
  - ◆ Soft drinks
  - ◆ Tea
  - ◆ Tobacco

INTERMEDIATE

▶ Explain the relationship to illness of:
  ◆ Water supply
  ◆ Sewage disposal
  ◆ Germ-bearing insects and pests

▶ Discuss cleanliness as protection from disease.

JUNIOR HIGH - HIGH SCHOOL

▶ Identify illnesses that result from poor nutrition.

  ⇒ **Primary deficiency** - Lack of nutrients
    ◆ Anemia
    ◆ Beriberi
    ◆ Goiter
    ◆ Night blindness
    ◆ Pellagra
    ◆ Rickets
    ◆ Scurvey

  ⇒ **Secondary deficiency** (body can't use the nutrient, causing a deficiency)
    Causes: vomiting, diarrhea, some medicines
    Consequences: possible dehydration, even starvation.

▶ Define obesity and identify illnesses it may lead to. (*diabetes, heart disease*)

▶ Offer a method of losing weight that will not be harmful to the body.

▶ A study of communicable diseases should include bacteria and viruses:
  ⇒ Bacteria: Define bacteria.
                 Explain the usefulness of bacteria.
                 Observe various shapes of bacteria.

  ⇒ Define pathogen.

  ⇒ Describe how diseases are transmitted. (insects, air)

⇒ Explain the role of antibiotics, medicines, and the body's immune system.

▶ Find out the causes and treatments of: diabetes, heart disease, mental illness.

| Activity | **Compare past and present approaches to medicine.**
What were considered causes and what were the treatments?

▶ Study of genetic disorders should include how they are acquired and by whom:
Inherited: Sickle-cell anemia, Tay-Sachs disease
Chromosome-related: Down's syndrome

▶ Identify illness related to the environment:
◆ Allergens and allergic reactions
◆ Chemicals (*lead poisoning*)
◆ Fiber floating in the air   (*lung disease common among coal miners*)
◆ Gases (*carbon monoxide poisoning and death*)

▶ Identify causes and treatments of the following illnesses:
◆ Auto-immune diseases
◆ Cardiovascular diseases (hypertension, stroke, paralysis)
◆ Cancer
◆ Multiple sclerosis

▶ Define and suggest treatment for sexually transmitted (venereal) diseases:
◆ Gonorrhea
◆ Hepatitis B
◆ Herpes simplex
◆ Syphilis
◆ Trichomoniasis

▶ Practice first aid, including:
⇒ Deciding what to do:
Stay calm
Check airway, breathing and heartbeat of the unconscious
Don't move the unconscious

⇒ How to get help

⇒ How to treat
◆ Bruises
◆ Burns
◆ Choking
◆ Drug overdoses
◆ Fractures

- ◆ Open wounds
- ◆ Sprains
- ◆ Stings

⇒ How to perform CPR (cardiopulmonary resuscitation).

⇒ How to perform mouth-to-mouth resuscitation.

DRUGS

▶ Define drug.

▶ Identify legal and illegal drugs.

▶ Give examples of proper and improper use of drugs.

▶ Identify drugs that may cause psychological dependence and those that may cause physical dependence.

▶ Explain withdrawal.

▶ Give the definition, short term and long term physical and mental effects, side effect, and results of misuse of:

Legal drugs:	Caffeine	(stimulant)
	Ethyl alcohol	(depressant)
	Nicotine (in tobacco) (stimulant)	

Illegal drugs:	Cocaine
	Hallucinogens:
	Hashish
	Marijuana (medical uses)
	LSD (lysergic acid diethylamide)
	PCP (phencyclidine)

Narcotics:	Heroin
	Legal by prescription
	Codeine
	Morphine

Prescription drugs:	Amphetamines	(stimulants)
	Barbiturates	(depressants)
	Tranquilizers	(depressants)

▶ Name sources of help for substance abuse.

EARTH SCIENCE

AIR AND WATER

weather, climate, atmosphere, solar system, oceans, seas

Weather

PRIMARY

▶ Observe and name different weather conditions:

   ◆ The elements of weather are: sunshine, temperature, moisture and wind.

   ◆ Describing weather requires more than one word:
      A morning could be described as cold, rainy, and windy.

▶ Terms to know:
   ◆ Cloudy
   ◆ Cool
   ◆ Cold
   ◆ Hot
   ◆ Humid
   ◆ Raining
   ◆ Snowing
   ◆ Still
   ◆ Sunny
   ◆ Warm
   ◆ Windy

▶ Observe that weather changes often.
A chart to record daily weather could use symbols for the different elements.
When the weather changes during the day, add more symbols.

▶ Identify clothing appropriate for different types of weather.
(Application of understanding gained from observations)

► Observe that while wind is air that is moving, we don't see the wind; we feel it or see something moving because of it.

► Observe the direction of the wind.
Hold up a pinwheel. Run while holding a stream of crepe paper.

► Observe that wind can blow hard or gently. Observe damage done by strong winds.

► Look for weather signs and make a prediction of coming or past weather.
Dark sky, big clouds: Rain
Rainbow, sun appearing: End of storm

 Have children recall the weather when they have seen a rainbow. Help them realize that drops of rain (rain or mist) must still be in the air when the sun shines. A rainbow is made when the sun shines on that drop and the colors in light are separated.

For Example:
Observe: puddles, wet sidewalk **Conclude**: rain recently ended
Observe: dry, cracked mud   **Conclude**: have had lots of dry, sunny days

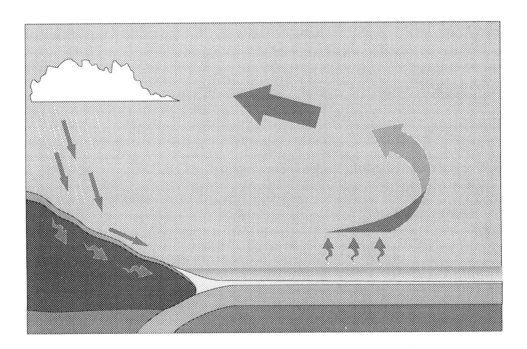

Experiences with evaporation and condensation are helpful before attempting to predict past weather (dried puddles) and future weather (humidity that may lead to rain). It provides a foundation for understanding the water cycle: water from the ground (lakes, oceans, etc.) evaporates, the water vapor condenses and returns as rain.

Activities

Evaporation

Place a dish with ¼ cup water on a window sill. Daily, measure the water in the dish until no more water is left. Where did it go? *(Tiny drops of water scatter in the air, but are so tiny we can't see them.)*

Condensation

(Cold touches warm—drops of water form.)

1. Place a jar of food from the cupboard (dry and at room temperature) into the refrigerator for a few hours to get cold. Take it out and let it sit at room temperature. How does the jar feel while taking it from the refrigerator? *(cold)* After sitting in the warm room? *(cold and wet)*

2. Boil water, observe steam rise, hold a cold pan over the steam and observe drops of water form.

▸ Observe and relate the readings of a thermometer to hot or cold.

Put a thermometer in a glass of hot water, observe, then put it in cold water and watch the liquid fall. At this stage children need only see that the colored liquid rises for hotter temperatures and lowers for colder temperatures.

▸ Observe characteristics of each season in the area in which the child lives.

INTERMEDIATE

▸ Recognize characteristics of seasons including changes in:
   ♦ Length of daylight
   ♦ Temperature
   ♦ Plants
      ⇒ Leaves that change color and then fall off
      ⇒ Formation of seeds
      ⇒ Flowers
   ♦ How animals deal with changing weather:
      ⇒ Hibernate
      ⇒ Change in coloring
      ⇒ Store food for winter
      ⇒ Migrate
      ⇒ Make a cocoon
      ⇒ Lay eggs

▸ Describe ways people prepare for various seasons.

▸ Observe that characteristics of a season vary from place to place.

▸ Read degrees on a thermometer after measuring temperature in Celsius and Fahrenheit.

▸ Find out the boiling point and freezing point of water.

▸ Realize that there are instruments used to measure:
   ◆ Wind speed   -   An anemometer
   ◆ Precipitation   -   A rain gauge

▸ Recognize that weather forecasts are based on a variety of records.
   (role of weather balloons, satellites)
   ◆ Air pressure: definition and causes
   ◆ High and low pressure areas
   ◆ Using a barometer to measure atmospheric pressure
   ◆ Cold and warm fronts: characteristics and how they are formed

▸ Define the following elements of weather:

Cloud	Made from water vapor. Many drops together (forming liquid water) make a cloud. Cloud types: cirrus, cumulus, stratus
Dew	Water vapor in the air near the ground condenses on the cool surfaces.
Fog	A cloud near the ground.

Precipitation:

Hail	Formed of many layers of ice—begin as raindrops.
Rain	Clouds become cool, the drops come together to make bigger drops (raindrops).
Sleet	Raindrops that freeze into ice as they fall.
Snow	Tiny bits of water vapor in the cloud become very cold and instead of becoming raindrops, become snowflakes (ice crystals).

⇒ Although **glacier** would be considered part of the land forms, this is an appropriate place to learn what it is: a river of ice fed by snow.

Familiarity with the water cycle, evaporation and condensation, and certain properties of air will increase understanding of the elements of weather.

▶ Explain the properties of air.
- ◆ Occupies space
- ◆ Has weight
- ◆ Exerts pressure (force pushing against an area)
- ◆ Cool air sinks
- ◆ Cool air contracts
- ◆ Warm air rises (pushed up by cool air)
- ◆ Warm air cools as it rises
- ◆ Warm air expands
- ◆ When warm air is cooled, the water vapor in it condenses.

Water cycle   The process of evaporation, condensation and rain is explained in terms of the movement of molecules, and properties of hot and cold air.

Evaporation   Molecules are always moving. Heating moves them faster (sunlight provides heat). Crowded molecules form a drop of water. When they move faster and separate they mix with the air.

Condensation   As air gets colder, it cannot hold as many molecules of water vapor. Heat is necessary for evaporation. When heat is removed, the water vapor will condense.

▶ At the intermediate level, students are introduced to definitions and causes of the following, as well as what to do for safety:
- ◆ Hurricanes
- ◆ Thunderstorms (lighting and thunder)
- ◆ Tornadoes

Weather and Climate

JUNIOR HIGH - HIGH SCHOOL

▶ Experiment to observe that air has pressure.

▶ Use a barometer to measure air pressure (called atmospheric pressure).

▸ Identify ways to raise or lower atmospheric pressure by decreasing or increasing temperature and humidity.

▸ Define:
  ◆ Dew point
  ◆ Humidity
  ◆ Relative humidity
  ◆ Saturated air

▸ Interpret weather by looking at a weather map:
  ◆ Air pressure
  ◆ Cold front
  ◆ Humidity
  ◆ Isobars
  ◆ Isotherms
  ◆ Temperature
  ◆ Warm front
  ◆ Wind

▸ Realize that water vapor in the air can be measured.

Chemical experiments can be conducted to show that the air does contain water vapor. Then experiments with condensation can be constructed in order to determine how much water is in the air.

▸ Relate the amount of water vapor in the air to:
existing temperature and air pressure.

▸ Define:
  ◆ Air mass
  ◆ Conduction
  ◆ Convection
  ◆ Coriolis effect
  ◆ Front
  ◆ Monsoon
  ◆ Prevailing winds: (Draw the prevailing winds on a map)
    ⇒ Doldrums
    ⇒ Trade Winds
    ⇒ Westerlies
    ⇒ Polar Easterlies

 Background should include an understanding of the interrelationship of the earth's rotation on its axis, wind, and the unequal heating of the surface of the earth by the sun.

▶ While looking at the heating of the earth's surface by the sun, study the various causes for uneven heating (diagrams or models are helpful):
⟹ Angle at which sun's rays strike the earth
⟹ Atmospheric conditions (moisture in the air...)
⟹ Latitude
⟹ Nearness of earth to sun
⟹ Season

▶ A study of air pressure and movement should lead to a more specific understanding of certain weather conditions:
⟹ Cyclone
⟹ Hurricane
⟹ Tornado

▶ Identify properties of air:
♦ Air has mass.
♦ Air has weight and exerts pressure.
♦ Air flows from areas of high pressure to areas of low pressure.

Concepts
• Energy is required for air to move (moving air is wind).
• Heat energy causes air to expand and become less dense.
• Warm, less dense air rises because of the greater pressure of the more dense, cold air with which it comes into contact.

▶ Explain the difference between weather and climate.

▶ Describe how various elements of weather are formed: hail, rain, sleet, snow
(Definitions are now more specific and include terms such as "freezing nuclei.")
▶ Give characteristics of various climate zones:

 Climate Zone labels vary. Check several maps to find the different labels. Those climates listed in biomes (see Animals) may be used and climate studied in relation to plant and animal life.

Atmosphere

JUNIOR HIGH - HIGH SCHOOL

▶ Define and assign a state to each (solid, liquid, gas) of the following, noting the continual interaction between these spheres.
  ◆ Atmosphere
  ◆ Hydrosphere
  ◆ Lithosphere

▶ Define biosphere.

▶ Name the layers of the atmosphere and characteristics of each, placing them in proper order.
  ◆ Exosphere
  ◆ Ionosphere
  ◆ Mesosphere
  ◆ Stratosphere
  ◆ Thermosphere
  ◆ Troposphere

▶ Define ozone layer.
  ⇒ Explain possible results if its concentration is reduced.

▶ Identify common air pollutants.
  Include the role of fluorocarbons, aerosol propellants, and so on, and the consequences people will experience as a result of too much ultraviolet radiation. This could be studied with a unit on health and the rising rates of skin cancers.

▶ Define magnetosphere.

▶ Describe the composition of air.

▶ Draw and label the nitrogen cycle.

▶ Draw and label the carbon dioxide cycle.

▶ Explain the "greenhouse effect."
  Discuss theories for and against the validity of this theory.

Solar System

planets, sun & stars, moon, outer space

PRIMARY

The young do best with concrete experiences. It is not necessary to cover this area at all at the primary level, but if you do, keep lessons simple and don't worry about retention.

▸ Concept: The earth rotates on an imaginary axis.

This concept should be taught in response to children's curiosity of what makes night and day. Use objects and a light source to illustrate your reply. Include the length of time for one full rotation.

▸ Concept: Gravity.

At this level it is explained simply as a pull of the earth: objects are pulled down so we don't fall off.

▸ A simple look at the stars should include:
  ⇒ The sun is a star.
  ⇒ Stars always shine.

  At the second or third grade level explanations may include:
  ⇒ What stars are made of.
  ⇒ The position of the sun in relation to the position of the sun's "height" in the sky, and it's usefulness for telling time (length and direction of shadow).

INTERMEDIATE - JUNIOR HIGH

Use picture books, models, and diagrams to build understanding of the following concepts.

▸ The earth revolves around the sun. Include:
  ♦ Length of time for one complete orbit.
  ♦ Shape of orbit (correctly labeled "ellipse").
  ♦ Distance of earth from sun.
  ♦ Position of earth in relation to other planets.
  ♦ Position of earth to sun in September, December, March, and June.

▸ The earth is part of a galaxy called The Milky Way.
  ⇒ Name the other planets in our solar system in the order of their orbits from closest to furthest from the sun.

⇒ Define the relationship between the length of time for one revolution around the sun and the distance from the sun.

⇒ Identify features of each planet in our solar system, including the length of time for one revolution on its axis, and one revolution around the sun.
Identify
- Shape of orbit (elliptical)
- Speed of the planet at each part of its orbit
- Length of time for one complete orbit

▸ Centrifugal and centripetal forces keep planets in orbit.
Demonstrate each of these forces.

▸ Identify several constellations, including the big and little dippers.
(*These may be studied along with an historical unit study of any ancient culture that used constellations to make predictions.*)

▸ Locate the North Star (Polaris).

▸ Use a telescope.
After experience using a telescope study:
⇒ Its usefulness
⇒ How it works
⇒ Solar telescope
⇒ Spectroscope

▸ Measure distance in light years.

▸ A study of the sun should include:
⇒ Identification of layers of the sun and characteristics of each:
- Chromosphere
- Corona
- Photosphere

⇒ Identification of the source of the sun's energy

⇒ Definition of fusion and how it is used to produce the sun's heat

⇒ Definition, causes, and effects of sunspots

⇒ Explanation of how the sun causes auroras on the earth

- A study of the moon should include:
    - Features of the moon
    - Why we see different shapes
    - Names of the phases of the moon
    - How a lunar eclipse occurs
    - The shape and length of time of its orbit

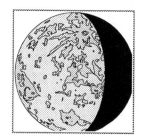

- Define or explain:
    - Comet (Haley's Comet; orbits; make-up of comets)
    - Galaxy
    - Gravity
    - Light year
    - Lunar eclipse
    - Mass
    - Weight
    - Meteor
    - Meteorite
    - Satellite
    - Shooting star
    - Solar eclipse
    - Solar system
    - Summer solstice
    - Vernal equinox
    - Winter solstice

- Study outer space including:
    - Date and names of the men and spacecraft that first landed on the moon
    - Uses of satellites and any current events relating to satellites
    - Voyages of spacecraft

INTERMEDIATE - HIGH SCHOOL

- Identify contributions to science by:
    - Nicolas Copernicus
    - Albert Einstein (including principles of relativity)
    - Galileo Galileo (including the telescope)
    - Johannes Kepler (laws derived by mathematics.)
    - Isaac Newton (include laws of gravitation and motion)

HIGH SCHOOL

Topics listed above continue at this level, but with greater depth.

▸ An explanation of seasons, equinox, and solstice should include the significance of the angles of the sun's rays:

⇒ The greater the slant, the less heat. (Energy is spread out over a larger area.)

⇒ At a right angle, light is brighter, and there is more heat. (Energy is concentrated.)

⇒ (The curve of earth's surface, and the tilt, and rotation of the earth affect how the sun's rays strike the earth.)

▸ Study the sun.   Include:

⇒ Composition (mostly hydrogen and helium)

⇒ Source of the sun's energy (Bethe's theory)

⇒ Radiant energy (a definition)

⇒ Spectroscope - used in analyzing light from stars (Used by astronomers to identify the elements that make up a particular star.)

⇒ Types of spectrums: continuous, bright-line, dark-line electromagnetic (Scientists can identify an element by its particular spectrum.)

▸ Identify several constellations.
Explain why some are found only in one season of the year.

▸ Explain why stars are different colors.

▸ Define:
   ◆ Black hole
   ◆ Doppler effect
   ◆ Galaxy
   ◆ Luminosity of a star
   ◆ Magnitude of stars
   ◆ Nebula
   ◆ Nova

▸ Study the moon.   Include:
   ◆ Spacecraft
   ◆ Requirements for landing a spacecraft
   ◆ Requirements (needs) in order to explore the moon

Water

INTERMEDIATE - JUNIOR HIGH

▶ Identify sources of water below the earth's surface:
  - ◆ Geyser
  - ◆ Spring
  - ◆ Water table
  - ◆ Well

▶ Define and explain the use of the following:
  - ◆ Filtering purifying plant
  - ◆ Reservoir
  - ◆ Watershed
  - ◆ Well

▶ Explain the causes and effects of water pollution and acid rain.
  (This may be studied along with food chains).

▶ A study of the ocean should include:
  - ⇒ Causes of tides (*Study this during a study of the moon.*)
  - ⇒ Patterns of current (*Look at direction on a world map*).
  - ⇒ Pattern of high and low tides
  - ⇒ Salinity
  - ⇒ Waves
    - ◆ Source
    - ◆ Measuring
    - ◆ Parts (crest, trough)
    - ◆ Tsunami (definition, causes)
  - ⇒ Ocean floor (for specifics see "The Land.")

HIGH SCHOOL
Continue with topics above, including:

▶ Causes of ocean currents: wind, temperature, differences in salinity

▶ The ocean is divided into zones. Describe characteristics of each zone.
  Abyssal Zone
  Bathyal Zone
  Neritic Zone
  Tidal Zone

The Land

composition, changes, use

PRIMARY

A study of the following concepts related to the composition of the earth could be covered while exploring plant and animal life and habitats.

▸ The earth is round.

▸ The earth's surface is covered with land and water.

▸ Land surface varies: mountains, hills, valleys, plains

▸ Water surface varies: oceans, seas, rivers, streams, lakes, ponds

▸ Rocks and soil are useful. (Identify common uses.)

INTERMEDIATE - JUNIOR HIGH

▸ Identify parts of the earth and the composition of each part:
- ◆ Core
- ◆ Mantle
- ◆ Crust

▸ Describe how a volcano is formed. (*Cause and effect of eruptions.*)

▸ Identify
- ◆ Cone
- ◆ Crater
- ◆ Lava
- ◆ Magma
- ◆ Vent

▶ Explain causes and effects of earthquakes.
  ⇒ Locate fault lines.
  ⇒ Explain how earthquakes may be predicted and measured.
  (*Earthquakes could be studied while looking at Japan to see how the people have adjusted their lives because of frequent quakes.*)

▶ Identify the make-up of three types of rocks, explaining how each type is formed:
Igneous
Metamorphic
Sedimentary

▶ Classify common rocks, placing them in one of the above categories.   Include:
Basalt
Granite
Limestone
Marble
Slate

▶ Identify common minerals. (*gold, silver, iron, copper*)

▶ Identify common minerals by physical properties:
  ◆ Cleavage
  ◆ Color
  ◆ Hardness
  ◆ Heft
  ◆ Luster
  ◆ Streak

▶ Identify properties of metal:
  ◆ Can change shapes without breaking by stretching (ductile) or hammering (malleable)
  ◆ Conducts heat well
  ◆ Distinct and shiny color (gold, copper)
  ◆ Hard (not easily chipped or scratched)

▶ Define weathering.
  Give examples of physical (mechanical) and chemical weathering.
  An explanation should include the cause.

Example:
The force of water weathers rock. Drops of water that have worked their way into the rock expand during freezing weather, cracking the rocks. Flowing water breaks down rock, carrying away bits of the rock (eroding it) as soil. A plant can act as a force in weathering as well, when its growth enlarges a crack in the rock. It causes chemical weathering if it produces acids that can crumble the rock.

- Identify the main causes of erosion:
  - Gravity
  - Wind
  - Water (flowing, waves, glaciers)

- Offer ways to prevent erosion.   (*CONSERVATION*)

Example: Plants help hold soil in place, so trees and ground cover are one solution. Farming by contouring fields to prevent rain from washing away top soil is another solution.

- Identify several sources of energy and how they are used.

Sources	*Used for*
Coal	Hauling (power)
Natural gas	Heat
Oil	Manufacturing
Sun	
Water	
Wind	
Wood	

JUNIOR HIGH - HIGH SCHOOL

- A look at the earth beneath the surface becomes more specific.
  (A drawing should either give distances of each section or a sense of the proportion for the size of each section.)
  - Inner Core
  - Outer Core
  - Mantle
  - Crust
    ⇒ Describe each layer.
    ⇒ Define bedrock.
    ⇒ Explain how scientists determined the make-up of the interior of the earth.

- Explain how caves may be formed.
  ⇒ Define stalagmite and stalactite.

- Define petrified wood, explaining how it is formed.

- Define fossil. Explain how fossils are formed and their usefulness to scientists.

▶ Explain how mountains are formed.
  ◆ Dome
  ◆ Fault-block
  ◆ Fold
  ◆ Volcanic

▶ Define (describe) the theory of plate tectonics.

▶ Define isostasy.

▶ Define
  ◆ Basin
  ◆ Dome mountains
  ◆ Hill
  ◆ Mesa
  ◆ Plain
  ◆ Plateau
  ◆ Valley

(Look at a relief map of the United States and locate mountain ranges and plains.)

▶ Describe the ocean floor:
  ◆ Abyssal plains
  ◆ Continental shelf
  ◆ Continental slope
  ◆ Mid-ocean ridge (and cause)
  ◆ Submarine canyons
  ◆ Trench

▶ Explain causes of earthquakes.
  ⇒ Identify major fault lines on a map or globe.
  ⇒ Explain how scientists determine the origin of an earthquake.
  ⇒ Explain how earthquakes are measured.
  ⇒ Terms to know:
    ◆ Epicenter
    ◆ Focus
    ◆ Pressure wave
    ◆ Richter Scale
    ◆ Seismic wave
    ◆ Seismograph
    ◆ Shear wave

▶ Diagram (or label a diagram) or explain the rock cycle.
(Magma becomes igneous rock, erosion produces sediment. Sediment may become sedimentary rock which can be turned into metamorphic rock, which can melt and become magma.)

▶ Explain how soil is formed.
  ⟹ Define
    ◆ Permeability of soil
    ◆ Porosity of soil
    ◆ Subsoil
    ◆ Topsoil
    ◆ Water table

  ⟹ Compare and contrast soils in various regions:
    ◆ Desert
    ◆ Forest
    ◆ Grassland
    ◆ Mountains
    ◆ Prairie
    ◆ Tundra

▶ Identify various uses of land. Look at the:
(This is often included in a look at methods of conservation.)
  ⟹ Percent useful for farming
    Percent not usable: desert, covered with snow, lacking topsoil

  ⟹ Factors hindering farming of the seas

  ⟹ Crop farming, including rotation of crops, methods of plowing to preserve topsoil

  ⟹ Trash disposal

  ⟹ Resources:
    ◆ Fresh water: power (electricity), irrigation, industry, human consumption
    ◆ Sea water: process for getting salt, fresh water, minerals, food
    Water pollution—chemical, sewage, oil spills, may be studied here.

▶ Identify sources of energy:
    ◆ Fossil Fuels: coal, gas, petroleum
        Where they are found
        How they are processed and used
    ◆ Geothermal energy
    ◆ Nuclear energy: fission, fusion
        Advantages and Disadvantages
    ◆ Sun (solar)
    ◆ Wind

▶ Define mineral.
    (Chemical elements or compounds found naturally.)

- ▶ Identify various minerals using properties of:
  - ◆ Cleavage
  - ◆ Color
  - ◆ Hardness
  - ◆ Luster
  - ◆ Specific gravity
  - ◆ Streak

 Chemical tests may be conducted to identify it by some other characteristic property. (Conduct flame tests for metallic elements. Refer to a chart indicating which element is associated with a particular color of flame. A magnet may help identify magnetite.)

- ▶ Identify crystal structures of various minerals.

- ▶ Find chemical names and formulas for a few minerals.

- ▶ Identify uses for various minerals.

- ▶ Classify several minerals as:
  - ◆ Metallic
  - ◆ Nonmetallic
  - ◆ Rock-forming

- ▶ Identify theories for determining the age of the earth.

- ▶ Investigate theories of how the world began.

 Public school textbooks often divide the earth into geologic periods covering millions of years and explain evolution as fact. Students should be given this information along with scientific rebuttals and evidence used to support the theory of creation so that they can examine the evidence on both sides.

PHYSICAL SCIENCE

FORCE

Motion, Work, Gravity, Magnetism

FORCE: Motion

A force is necessary to make things move.

PRIMARY

Guide children toward discoveries about how things move:

▸ A force is needed to start, stop, or change the direction of motion of an object.

▸ A force can be a push or pull on an object.

▸ Different kinds of force can be used:
   ◆ Muscles (of animals and of humans)
   ◆ Moving air
   ◆ Moving water

▸ It takes more force to move heavy things than light things.

▸ It takes less force to move an object if wheels are used.
   (*Compare using muscles to push or pull an object first without, then with, wheels.*)

INTERMEDIATE - JUNIOR HIGH

Use experiences and experiments to teach these concepts:

▸ Newton's Laws of Motion:

1. Anything at rest (not in motion) will stay at rest unless a force is applied (to make it move.) This is called INERTIA

⇒ When one object touches another to move it, **direct force** has been applied.
   Gravity is an **indirect force** that moves an object.

---

2. Anything in motion will stay in motion (forever) in a straight line with unchanging speed unless a force changes its motion.

⇒ Change in motion is the result of energy applied to produce an unbalanced force.

⇒ If all forces on an object are balanced, there will not be a change in motion.

▶ Force can be applied in one of two ways: pushing or pulling
(Lifting is either pushing up or pulling up).

▶ When force is applied to an object (a mass), it accelerates.

▶ Pressure (a pushing force) equals force divided by area.

▶ Force can be measured. (Use a scale with newtons as units.)

▶ Different kinds of force can be used to make things move:
   ◆ Electricity (See "Energy.")
   ◆ Magnetism (See "Magnetism.")
   ◆ Moving water
   ◆ Muscles (humans, animals)
   ◆ Springs
   ◆ Wind (moving air)

▶ Forces can be combined.

▶ Define friction.
   ⇒ Describe its usefulness. *(There is little or no friction in outer space.)*
   ⇒ Discover how friction affects the force needed to move an object. *(More force is needed.)*

HIGH SCHOOL
The concepts above continue to be studied while investigating forces of gravity, work, magnetism, and energy. Exercises should include measuring and calculating.

▶ Students should understand Newton's three Laws of Motion:

1. The **Law of Inertia** explains why a force is needed. If an object is at rest it remains at rest. If an object is in motion it remains in motion, moving in a straight line and at a constant speed UNLESS acted upon by unbalanced forces.

⇒ Friction must be overcome for an object to slide over another object.

⇒ Gravity must be overcome for an object to be lifted.

2. The second law explains why things speed up or slow down:

⇒ The force acting on an object, the object's mass, and the change in speed that results are related.

⇒ Applying a small force to an object of large mass makes a small change in speed.

⇒ Applying a great force to an object of small mass makes a great change in speed.

⇒ The force can push or pull in the same direction, increasing speed.

⇒ The force can push or pull in the opposite direction, decreasing speed.

3. Action equals Reaction.
For every action force there is an equal and opposite reaction force.

⇒ When one object causes a force to act on a second object, that second object causes an equal force to act on the first object, but in the opposite direction.

FORCE: Gravity

INTERMEDIATE - JUNIOR HIGH

▸ Identify the law of gravity and understand gravity as a force.
The force of gravity is a pull. Large objects attract (pull toward themselves) smaller objects.

▸ Define centrifugal force.
⇒ Explain how it and gravity keep the moon in an orbit around the earth.

JUNIOR HIGH - HIGH SCHOOL

▸ Explain why objects fall using Newton's idea of gravitational attraction.

▸ Explain the acceleration of falling objects.

▸ Define terminal speed.

▸ Define mass and weight in terms of gravitational attraction and explain the difference between the two.

▸ Calculate using the formula: Density = Mass divided by Volume

FORCE: Work

HIGH SCHOOL

▸ Give a scientific definition of work and power.

▸ Calculate using the following formulas:

Work = Force x Distance
Effort = Resistance divided by M.A.
M.A. = Length divided by Height (inclined plane)
Distance = Speed x Time
Power = (Force x Distance) divided by Time

FORCE: Work Using Machines

Machines make man's work easier.

PRIMARY

▸ Identify common objects as simple machines. (Ask how they make a job easier.) hammer, saw, screwdriver, bottle opener, lawnmower, blender, and so on.

▸ Identify the force that moves the simple machines.

INTERMEDIATE

▸ Identify simple machines:
  ◆ Inclined plane
  ◆ Lever
  ◆ Pulley
  ◆ Screw
  ◆ Wheel and axle
  ◆ Wedge

 Simple machines change the amount of applied force and/or the direction of applied force. They can change the direction of movement.

JUNIOR HIGH - HIGH SCHOOL

▸ Identify characteristics of machines:

⇒ Used to overcome forces of gravitation, inertia, friction.

⇒ Increase force, increase speed, change the direction of force, transfer force.

 The amount of energy gotten out of a machine does not exceed the energy put into it.

▸ Explain the mechanical advantage of using machines, using the terms *effort* and *resistance*.

SIMPLE MACHINE: Wheel and Axle

PRIMARY - INTERMEDIATE

Conduct experiments to discover the following concepts and understand how wheels, axles and pulleys work.

⇒ Round things roll.

⇒ There are different kinds of wheels.

⇒ Wheels need axles to turn.
A pulley is one kind of wheel; it turns on an axle. A rope fits around the pulley. When one end of the rope is pulled, the other end, and what is attached to it, moves.

⇒ Wheels have many uses:
  ◆ Some roll from place to place. (wheels on wagons, cars, bicycles)
  ◆ Some turn but do not move. (doorknob, knobs on stove, wheel on can opener)
  ◆ Some wheels can make other wheels turn with a belt, chain, or interlocking teeth.

 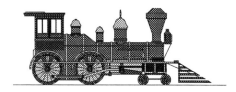 

SIMPLE MACHINE: Inclined Plane

PRIMARY - INTERMEDIATE

▶ Discover that ramps (inclined planes) make it easier to move things.

JUNIOR - HIGH SCHOOL

▶ Define effort and resistance.

▶ Explain why the longer the distance over which the object moves (effort distance) the greater the mechanical advantage. (*Resistance distance is the height from the ground to the highest part of the inclined plane.*)

▶ Explain how a screw is a kind of inclined plane. (*an inclined plane wrapped around something*)

▶ Explain how a wedge is like an inclined plane (shape) and how it is different. (*You move it, rather than the object.*)

SIMPLE MACHINE: Levers

PRIMARY - INTERMEDIATE

▶ Discover levers make work easier.
Words to know:
Force arm
Weight arm
Fulcrum

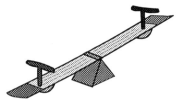

▶ Discover levers can be used as a balance.

 The position of the fulcrum affects the amount of force needed to lift an object, and how high it can be lifted. (The nearer the fulcrum is to the weight, the less the force is needed to lift the weight; the nearer the fulcrum is to the force, the higher the weight can be lifted.)

JUNIOR HIGH - HIGH SCHOOL

▶ Define the categories
  ◆ First-class
  ◆ Second-class
  ◆ Third-class.

▶ Identify various objects as a first, second, or third class lever.

FORCE: Water

PRIMARY

▶ Discover objects that float. (*Water is the force pushing up.*)

▶ Discover how water affects the weight of an object:
When does an object feel heavier—when it's being held under water, or in the air?
(*Air*)

▶ Lift an object from under the water, out of the water. Does the object seem to gain weight?

▶ Discover water as a force for machines: water wheel

INTERMEDIATE - HIGH SCHOOL

 The weight of an object changes when it is immersed in a liquid.

▶ Experiment to find out if things weigh more or less in water.
If you uses a spring scale you can attach the object (washers work well) and hold it in the air and record the weight. Objects should weigh less in water.

 Objects float because of buoyant force (the upward force).
(*An object must displace (push aside) its own weight of water, or more, in order to float.*)

▶ Experiment to discover what objects float.
Make boats and find out which designs float the best.

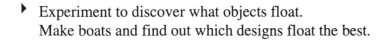

▸ Develop an understanding of the following concepts while studying buoyant force:
- Volume
- Mass
- Weight (realize weight may change, mass doesn't)
- Displacement of water

▸ Calculate volume.

FORCE: Air

PRIMARY

▸ Observe that moving air is a force. (*wind*)

(If a student puts his hand out the window while riding in a car, he can feel the force of air against his palm.)

▸ Discover that air can be a useful force. (*windmill*)

INTERMEDIATE

▸ Experiment to understand the following concepts:

♦ Air is made up of several gases.

♦ Air provides a force of push when it is moving, and when it is not moving.

♦ The amount of push depends on the amount of air.

♦ Escaping air can be a force.

Blow up a balloon, then let go—watching the <u>escaping</u> air push the balloon through the air.

INTERMEDIATE - JUNIOR HIGH

▶ Discover what a magnet does and does not attract.

▶ Identify poles on a magnet.

▶ Conduct experiments that lead to the following understandings:
   ◆ A magnet's size and its force are not necessarily related.
   ◆ A magnet has a magnetic field. (Use a magnet and iron filings)

▶ Magnetize another object.

▶ Find out how a compass works. (Including the understanding that the earth has a magnetic north and south pole which is different than the poles used in geography.)

HIGH SCHOOL

▶ Define lodestone.

▶ Identify properties of magnets:
   ◆ Like poles repel.
   ◆ Opposite poles attract.
     (Bar magnets have a north pole at one end and south pole at the other.)
   ◆ A magnet can be used to magnetize other pieces of iron.
   ◆ A magnet may be broken into pieces, each piece will be magnetic.

▶ Define magnetic field.

The following ideas are often taught under the double heading of electricity and magnetism:

▶ Identify an electromagnet and its use.
   (Many texts and books of experiments include making an electromagnet.)

▶ Explain Faraday's contribution to science by using magnets to make electricity.

▶ Explain what a transformer is and its use.

ENERGY: Heat Energy

PRIMARY

▶ Discover sources of heat:
- ◆ Sun (cooler in shade, clouds cut off some heat)
- ◆ Fire (needs air to burn)
- ◆ Electricity

▶ Identify many uses for heat. (*dry clothes, cook food, keep us warm*)

▶ Discover different ways for things to get hot:
- ◆ Heat moves through some materials (metal) but not others (wood, cloth).
- ◆ When air is heated it moves and gives it to whatever it touches. (A radiator or stove heats a room, for instance.)
- ◆ When air is heated it moves upward.

INTERMEDIATE - JUNIOR HIGH

▶ Identify low temperature with slow moving molecules and high temperature with fast moving molecules.

▶ Explain (describing the action of the molecules) why things expand when heated and contract when cool.

▶ Define three ways heat moves from hot objects to cold objects:
- ◆ Conduction
- ◆ Convection
- ◆ Radiation

▶ Explain how heat energy can cause changes in matter:
- ◆ Through transfer of heat. (*A cup of hot soup left at room temperature would cool because the heat is being transferred to the surroundings.*)
- ◆ Through expansion or contraction. (*Change in state—solid to liquid, etc.*)

▶ Identify a material as a conductor or insulator according to how easily energy travels through the material.

▶ Explain how friction can produce heat:
The energy of motion is transformed into heat.

JUNIOR HIGH - HIGH SCHOOL

▶ Define energy.

▶ Explain the difference between potential and kinetic energy.

▶ Explain the kinetic theory of matter (see Chemistry), heat transfer, contraction and expansion.

▶ Explain the law of the conservation of energy.

▶ Identify several sources of heat and describe how heat is produced in each:
  ◆ Chemical reactions
  ◆ Friction
  ◆ Heat from fuels (combustion)
  ◆ Nuclear reactions
  ◆ Solar energy

▶ Explain how heat is produced in each of the following.
  ◆ Atomic heat
  ◆ Combustion
  ◆ Electricity

▶ Explain how heat travels. (*convection, conduction, radiation*)
  Include the role of insulation.

▶ Define engine.
  (*A machine that changes heat energy into mechanical energy.*)

▶ Explain the difference between an internal and external combustion engine.
  (This could be included with a study of machines.)

▶ Explain the differences between heat and temperature.
  (*Understand that atoms move faster when heated.*)

▶ Explain how a thermometer works.
  Use the terms:
  ◆ Molecules expansion
  ◆ Contraction

▶ Identify ways of measuring temperature and heat:
  ◆ Instruments (*calorimeter - for example*)
  ◆ Units   (*calorie*)

▸ Identify ways heat affects matter:
(*This may be included with a study of chemistry or weather.*)
- ◆ Raises temperature
- ◆ Expands volume
- ◆ Decreases density
- ◆ Changes state (*solid to liquid, liquid to gas*)

ENERGY: Radiant Energy

See EYE and LIGHT, also.

JUNIOR HIGH - HIGH SCHOOL

▸ Define and name forms of radiant energy.

▸ Define electromagnetism and electromagnetic spectrum.

▸ Define radiation.

▸ List pros and cons of:
- ◆ X-rays
- ◆ Radiation from nuclear reactions
- ◆ Ultraviolet rays

▸ Define (*Use in study of eye and light*):
- ◆ Angle of incidence
- ◆ Angle of reflection
- ◆ Concave mirror
- ◆ Convex mirror
- ◆ Converging light rays
- ◆ Diverging light rays
- ◆ Farsighted
- ◆ Nearsighted
- ◆ Real image
- ◆ Virtual image

▸ Define:
- ◆ Luminous
- ◆ Reflected light
- ◆ Refracted light

► Explain how an electric light works.

Terms:
- ♦ Bulb
- ♦ Filament
- ♦ Florescent light
- ♦ Incandescent light
- ♦ Nitrogen gas
- ♦ Tungsten

ENERGY: Electricity

INTERMEDIATE - JUNIOR HIGH

► Discover the path of electricity using a dry cell battery, wires, electric light bulb and socket; get the bulb to light. (*It travels in a complete circuit.*)

JUNIOR HIGH - HIGH SCHOOL

► Define static electricity.

► Identify materials as good conductors and tell why.

► Identify materials as good insulators and tell why.

► Terms to know:
- ♦ Electron
- ♦ Grounded
- ♦ Negative charge
- ♦ Neutral
- ♦ Positive charge

 Concept: Unlike charges attract, like charges repel.

► Define voltage.

► Define:
- ♦ Alternating current
- ♦ Circuit
- ♦ Direct current
- ♦ Electric current

▶ Explain the difference between a series circuit and a parallel circuit.
Explain how fuses and circuit breakers protect electric circuits.

▶ Identify instruments and units of measurement:
   ◆ Ampmeter
   ◆ Ampere (amp)
   ◆ Ohm
   ◆ Volt
   ◆ Voltmeter
   ◆ Watt

▶ Calculate using the following formulas:
   ◆ Voltage x Amperage = Power (in watts)
   ◆ Voltage divided by Amperage = Resistance (in ohms)

▶ Define
   ◆ Chemical electricity
   ◆ Dry cell
   ◆ Electrodes
   ◆ Electrolyte

▶ Explain what electrolysis does to water.

▶ Explain the process of electroplating metals.

▶ Identify sources of energy that could be used to produce electricity.
Offer pros and cons.

ENERGY: Waves

See EAR AND SOUND, also.

JUNIOR HIGH - HIGH SCHOOL

▶ Define:
   ◆ Frequency - hertz
   ◆ Wave   (Realize that a wave moves energy from one place to another.)
   ◆ Wavelength
   ◆ Longitudinal wave (Including compressions and rarefactions)
   ◆ Transverse wave

▶ Label points of a wave:
  ◆ Amplitude
  ◆ Crest
  ◆ Trough

▶ Calculate speed based on the formula: Frequency x Wavelength = Speed

CHEMISTRY: Matter

Students should conduct experiments to acquire understanding of concepts.

PRIMARY

 Concept: Matter exists as solids, liquids, and gases.

▶ Students should observe properties of solids, liquids and gases:
  ◆ Solids keep their shape.
  ◆ Liquids take the shape of the container.
  ◆ They can't see gases.

▶ Explain the following properties of matter:
  ◆ Has weight (*on earth, since there is gravity, but at this level only having weight is likely to be understood.*)
  ◆ Takes up space.
  ◆ Is made up of molecules.

 That matter can change states if heat (energy) is added or removed may be discovered by using simple experiments. At this level the idea of heat energy may be too difficult to understand. Rather, water is frozen and they see an ice cube, water is boiled and they see steam.

INTERMEDIATE

Concepts listed under Junior and Senior High may be introduced at this level. Students should conduct experiments or make observations that will lead to understanding, rather than merely memorizing concepts.

▶ Define properties of matter:
  ♦ Density (the more mass in a given volume, the higher the density)
  ♦ Volume (takes up space)
  ♦ Mass
  ♦ Weight (where there is gravity)

▶ Experiment to find the role on various substances of:
  ♦ Heat
  ♦ Pressure
  ♦ Temperature

▶ Define:
  ♦ Density
  ♦ Heat
  ♦ Mass
  ♦ Pressure
  ♦ Specific gravity
  ♦ Temperature
  ♦ Vacuum
  ♦ Volume
  ♦ Weight

▶ Explain the difference between mass and weight.
  (*Matter always has mass, but doesn't always have weight. For example, in outer space there is no pull of gravity so matter floats and is weightless.*)

▶ Determine the following for various substances:
  ♦ Mass
  ♦ Volume
  ♦ Density
  ♦ Weight
  ♦ Temperature

▶ Calculate air pressure with a barometer.

▶ Discover why objects float or sink:
  ♦ Buoyancy
  ♦ Density
  ♦ Displaced water
  ♦ Volume

▶ Explain the kinetic theory of matter. Include the following ideas:
- ◆ Molecules are in constant motion. (This motion is heat energy.)
- ◆ Gas molecules move faster than liquid molecules at the same temperature.
- ◆ Molecules in a solid have only limited motion.
- ◆ The motion of molecules depends on their temperature.
  (Add heat, molecules move more rapidly and possess more kinetic energy.)

▶ Students should learn the follows concepts:

⟹ Molecules are made up of atoms and each kind of matter has its own molecule. Example: a molecule of water is different than a molecule of sugar.

⟹ There is space between molecules but a force (a pulling together) keeps them together.

⟹ In the solid state molecules are close together and vibrate, keeping their positions. (The molecules are closer together and move more slowly than in a liquid or gas.)

⟹ Solids are divided into two groups: amorphous solids and crystals.

⟹ In the liquid state, the molecules move around, but not far from each other. (They are more loosely packed than in a solid.)

⟹ The molecules bounce around and are far apart in gases.

⟹ Matter can be changed from one state to another by adding or removing heat energy. (solid to liquid, liquid to gas, and so on)

⟹ Different amounts of heat energy are required for different kinds of molecules. (You can melt chocolate in a pan but the pan doesn't melt.)

⟹ A physical change in the state of matter does not change the structure of the molecules. Rather, it changes the motion of the molecules. (melting, freezing, boiling, cutting, bending, squeezing, sifting) The total amount of matter does not change.

⟹ In a chemical change atoms react and there is a change in the molecules. (The total amount of matter does not change.)

▶ Find examples of solids with any of the following properties:
- ◆ Brittleness
- ◆ Ductility
- ◆ Elasticity
- ◆ Hardness
- ◆ Malleability

▸ Define tensile strength.

▸ Experiment to see how temperature may influence viscosity.
   Viscosity is a property of liquids—whether they flow easily or not.

▸ Surface tension is the "skin" formed on the surface of a liquid.
   Find out why this happens.

CHEMISTRY: Atoms, Elements, and Compounds

Some of the most basic concepts can be introduced at the intermediate level.

JUNIOR HIGH - HIGH SCHOOL

▸ Identify the structure of an atom, labeling parts and giving characteristics of each:
   ♦ Electron
   ♦ Neutron
   ♦ Nucleus
   ♦ Proton

▸ Define:
   ♦ Lepton
   ♦ Quark

▸ Students should learn the following concepts:

   ⇒ Each atom has at least one electron and one proton.

   ⇒ The number of electrons is the same as the number of protons.

   ⇒ Every atom except hydrogen has neutrons.

   ⇒ Electrons move in an orbit and have a negative charge.

   ⇒ In the nucleus: the proton has a positive charge, the neutron is neutral (no charge).

Note: During the study of cells, include nucleus as the *center* that makes it possible for cells to divide and reproduce.

▸ Identify illustrations of the atoms of hydrogen and helium.

▶ Define molecule. (Include: atoms share electrons).

▶ Define element.

▶ Explain how elements are arranged on the *Periodic Table of Elements*, and why.

▶ Each element has its own symbol on the Periodic Table.

⇒ Identify symbols for common elements such as gold, silver, oxygen.

⇒ Identify and define the mass number (sum of protons and neutrons in the atom), and the atomic number (number of protons).

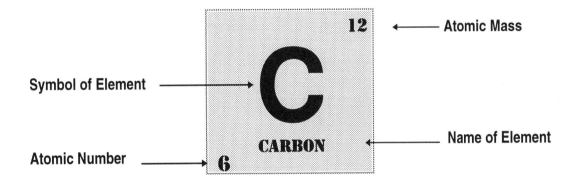

▶ Define isotope.

▶ Draw models of several elements showing the protons, neutrons and electrons in orbit.

▶ Elements are divided into two categories: metals and nonmetals.
List characteristics of each category and identify several elements as a metal or a nonmetal. (See *Solids: Metals, Nonmetals, Crystals*)

▶ Write the symbols for a variety of common compounds.
(H_2O for water, CO_2 for carbon dioxide, and so on.)

▶ Explain the meaning of the symbols and numbers for various compounds.
Example: H_2O: A molecule of water consists of 2 hydrogen atoms and 1 oxygen atom.

▶ Diagram a variety of elements and compounds showing the nucleus, the number of protons and neutrons, and the electrons in orbit.

▶ Write a structural formula for several compounds that demonstrate bonding. (H-O-H for water shows the oxygen atom sharing bonds with hydrogen.)

▶ Write the chemical equation for photosynthesis.

▶ Write chemical equations.
Include writing the formula during experiments or studies that explain a chemical reaction taking place.

▶ Define mixture and explain how it is different than a compound.
(Elements mixed are unchanged, but in a compound the elements are changed into a new substance.)

▶ Explain the difference between an homogeneous and heterogeneous mixture.
(Homogeneous mixtures are also called solutions.)

▶ Describe ways of separating mixtures.
   ◆ Distillation for some liquids (Uses the differences in boiling points.)
   ◆ A centrifuge machine for blood
   ◆ A magnet for solids — steel from brass

▶ Define compound.
Include: combination of two or more different kinds of atoms.
   ◆ Atoms are attracted to each other by a force called a chemical bond.
   ◆ Some bonds are stronger than others.
   ◆ A compound is different from the elements that it is made of.

Example: Gases hydrogen and oxygen make up a liquid — water.

▶ Define:
   ◆ Chemical bond
   ◆ Covalent bond
   ◆ Covalent compound
   ◆ Ion
   ◆ Ionic bond
   ◆ Ionic compound

▶ Conduct experiments to see how a compound is different from its elements.

▶ Observe different types of chemical changes:
   ◆ Combining elements into a compound
   ◆ Separating compounds into elements
   ◆ Rearranging compounds into new compounds

▶ Make solutions.

▶ Describe the strength of a solution: percentage by volume.
  Example: A 5% by volume solution of salt water would be 5/100 salt and 95/100 water.

▶ Define:
  ◆ Alloy
  ◆ Aqueous solution
  ◆ Colloid
  ◆ Solution
  ◆ Solid solution
  ◆ Solvent
  ◆ Suspension

▶ Explain *dissolve*:
  ◆ How substances dissolve (molecular attraction, energy involved)
  ◆ Rate of solution.

▶ Define (explain):
  ◆ Solubility (concentrated and dilute)
  ◆ Saturated, supersaturated, unsaturated solutions
  ◆ Boiling and freezing points

▶ Prepare dilute, saturated, and supersaturated solutions.

▶ Explain the difference between a solution and a suspension.

▶ Conduct tests to identify an unknown substance by finding properties that would identify it with a group of known substances.

▶ And/Or Identify common household matter as acid or base.
  (fruit juice, vinegar, shampoo, baking soda)

▶ Test for acids.

▶ Test for bases (alkalis).
  ⇒ Indicators to use for experimenting: BTB (bromthymol blue), litmus, methyl orange.
  (*Indicators are matter that have different properties in acids than they do in bases.*)

▶ If a substance is neither an acid or base, it is neutral.
  Find examples of neutral substances.

▶ Test for starch.

At the intermediate and junior high level a common property of acids is given as "sour taste" and for a base, "bitter taste." However, be sure that students are aware that they should never test an unknown substance by taste!

HIGH SCHOOL

▶ Characteristics of acids and bases may be described according to what an acidic or basic solution can do:
   ◆ Both conduct an electric current.

   ◆ Acidic solutions also neutralize bases, turn litmus solutions red, and produce hydrogen gas as a reaction with certain metals.

   ◆ Basic solutions also neutralize acids, turn litmus solution blue, and feel slippery when touched. (Be sure no one touches an unknown chemical!)

 Concept: Every element gives off its own color of light when heated. This can be seen in an instrument called a spectroscope.

▶ Draw structural formulas.

▶ Interpret and write numerous chemical formulas.

CHEMISTRY: Gas

Simple concepts may be introduced at the Intermediate Level.

JUNIOR HIGH - HIGH SCHOOL

▶ Conduct experiments that will increase understanding of behavior of gases.

 The kinetic theory of gases:
   ◆ Gases are made up of single atoms or molecules which are in constant motion.
   ◆ There is a great deal of space between the atoms or molecules.
   ◆ When gas particles collide with something (each other, walls of a container), they rebound without losing energy.
   ◆ The speed of the moving particles increases with temperature.

Pressure and Volume

 Boyle's Law: As the volume of a gas is decreased, the pressure of the gas increases (when the temperature remains constant).

Temperature and Volume

 Charles' Law: When the pressure remains unchanged, the volume of a gas increases as the temperature rises, or decreases as the temperature falls.

Oxygen cycle: Oxidation   iron - rust

CHEMISTRY: Water

INTERMEDIATE - HIGH SCHOOL

▶ Study to understand the following concepts about water:

 ◆ In the solid state: ice
   Ice is less dense than in its liquid state. (This is the opposite of the usual case.)
   Therefore, ice forms at the top of a pond and doesn't sink to the bottom.

 ◆ In the liquid state: water is a solvent (substances dissolve in it).

 ◆ Water exerts pressure.
   This pressure increases with depth.
   The pressure it exerts upward is its buoyant force (which makes things float).

 Archimedes' Principle: The buoyant force (upward push) of the water on an object in the water is equal to the weight of the water which was displaced (pushed aside) by the object.

 ◆ In the gas state:
   The boiling point of water is 212 degrees Fahrenheit. The temperature stays here until the water changes to steam (water vapor).

 Triple Point: The one pressure and temperature at which all three states (solid, liquid, gas) can exist at the same time.

INTERMEDIATE - HIGH SCHOOL

▶ Experiment and study to learn about the following properties of metals:
   ♦ Conduct electricity well.
   ♦ Conduct heat well.
   ♦ Are always shiny on the inside (sometimes on the outside too).
   ♦ Each metal has its own characteristic color.
   ♦ Can be changed into different shapes without cracking or breaking:
      ♦ By hammering = malleable
      ♦ By stretching = ductile

▶ Define alloy and give examples:

Alloy: A mixture of metals or of metal and nonmetals.
Brass: An alloy of copper and zinc
Bronze: An alloy of copper and tin.
Steel: An alloy of iron and carbon (a metal and a nonmetal)

▶ Define nonmetals:
They do not have the properties of metal, although some may conduct heat.
Examples: Elements such as carbon and oxygen
Compounds such as wood, rocks, glass.

▶ Define and observe crystals.
Both metals and nonmetals can be in the form of crystals.
   ♦ Crystals have repeated regular designs.
   ♦ Metal crystals have the properties of metals.
   ♦ Nonmetal crystals have the properties of nonmetals.

PERIODIC TABLE OF ELEMENTS

PERIODIC TABLE OF ELEMENTS

Elements 1-2, 7-10, 17-18, 36, 54, and 86 are gases at room temperature (20°C).

Elements 35, and 80 are liquids at room temperature (20°C).

Elements 43, 61, and 93-109 are produced in laboratories and do not occur in nature.

All other elements listed are solids at room temperature.

(Students may find it helpful to color-code the table.)

1	2	3	4	5	6	7	8	9
1 1.0 **H** Hydrogen								
3 6.9 **Li** Lithium	**4** 9.0 **Be** Beryllium							
11 22.9 **Na** Sodium	**12** 24.3 **Mg** Magnesium							
19 39.1 **K** Potassium	**20** 40.0 **Ca** Calcium	**21** 44.9 **Sc** Scandium	**22** 47.9 **Ti** Titanium	**23** 50.9 **V** Vanadium	**24** 51.9 **Cr** Chromium	**25** 54.9 **Mn** Manganese	**26** 55.8 **Fe** Iron	**27** 58.9 **Co** Cobalt
37 85.4 **Rb** Rubidium	**38** 87.6 **Sr** Strontium	**39** 88.9 **Y** Yttrium	**40** 91.2 **Zr** Zirconium	**41** 92.9 **Nb** Niobium	**42** 95.9 **Mo** **Molybdenum**	**43** 98.9 **Tc** Technetium	**44** 101.0 **Ru** Ruthenium	**45** 102.9 **Rh** Rhodium
55 132.9 **Cs** Cesium	**56** 137.3 **Ba** Barium	**La - Lu ***	**72** 178.4 **Hf** Hafnium	**73** 180.9 **Ta** Tantalum	**74** 183.8 **W** Tungsten	**75** 186.2 **Re** Rhenium	**76** 190.2 **Os** Osmium	**77** 192.2 **Ir** Iridium
87 223.0 **Fr** Francium	**88** 226.0 **Ra** Radium	**Ac - Lr ****	**104** 261.1 **Unq** Unnilquadium	**105** 262.1 **Unp** Unnilpentium	**106** 263.1 **Unh** Unnilhexium	**107** 262.1 **Uns** Unnilseptium	**108** **Uno** Not Named	**109** **Une** Not Named

* La - Lu

57 138.9 **La** Lanthanum	58 140.1 **Ce** Cerium	59 140.9 **Pr** Praseodymium	60 144.2 **Nd** Neodymium	61 146.9 **Pm** Promethium	62 150.3 **Sm** Samarium	63 151.9 **Eu** Europius

** Ac - Lr

89 227.0 **Ac** Actinium	90 232.0 **Th** Thorium	91 231.0 **Pa** Protactinium	92 238.0 **U** Uranium	93 237.0 **Np** Neptunium	94 244.0 **Pu** Plutonium	95 243.0 **Am** Americius

Legend:

Atomic Number	6	12.0	Atomic Mass
	C		Symbol of Element
	Carbon		Name of Element

10	11	12	13	14	15	16	17	18
								2 4.0 **He** Helium
			5 10.8 **B** Boron	6 12.0 **C** Carbon	7 14.0 **N** Nitrogen	8 15.9 **O** Oxygen	9 18.9 **F** Fluorine	10 20.1 **Ne** Neon
			13 26.9 **Al** Aluminum	14 28.0 **Si** Silicon	15 30.9 **P** Phosphorus	16 32.0 **S** Sulfur	17 35.4 **Cl** Chlorine	18 39.9 **Ar** Argon
28 58.7 **Ni** Nickel	29 63.5 **Cu** Copper	30 65.3 **Zn** Zinc	31 69.7 **Ga** Gallium	32 72.5 **Ge** Germanium	33 74.9 **As** Arsenic	34 78.9 **Se** Selenium	35 79.9 **Br** Bromine	36 83.8 **Kr** Krypton
46 106.4 **Pd** Palladium	47 107.8 **Ag** Silver	48 112.4 **Cd** Cadmium	49 114.8 **In** Indium	50 118.6 **Sn** Tin	51 121.7 **Sb** Antimony	52 127.6 **Te** Tellurium	53 126.9 **I** Iodine	54 131.3 **Xe** Xenon
78 195.0 **Pt** Platinum	79 196.9 **Au** Gold	80 200.5 **Hg** Mercury	81 204.3 **Tl** Thallium	82 207.1 **Pb** Lead	83 208.9 **Bi** Bismuth	84 209.9 **Po** Polonium	85 208.9 **At** Astatine	86 222.0 **Rn** Radon

64 157.2 **Gd** Gadolinium	65 158.9 **Tb** Terbium	66 162.5 **Dy** Dysprosium	67 164.9 **Ho** Holmium	68 167.2 **Er** Erbium	69 168.9 **Tm** Thulium	70 173.0 **Yb** Ytterbium	71 174.9 **Lu** Lutetium
96 247.0 **Cm** Curium	97 247.0 **Bk** Berkelium	98 251.0 **Cf** Californium	99 252.0 **Es** Einsteinium	100 257.0 **Fm** Fermium	101 258.0 **Md** Mendelevium	102 259.1 **No** Nobelium	103 260.1 **Lr** Lawrencium

TECHNOLOGY: Electronics

JUNIOR HIGH - HIGH SCHOOL

▶ Define electronics.

▶ Describe how electrons can be freed from atoms in a metal (*Edison effect*) and why that is important. (*They can then flow as an electric current.*)

▶ Find out how the following work: radio, television, computer.

TECHNOLOGY: Engineering

JUNIOR HIGH - HIGH SCHOOL

▶ Explain the difference between scientists and engineers.

▶ Explain the difference between:
civil, mechanical, chemical, and electrical engineers.

 A study of bridges would fit into this category.

TECHNOLOGY: Flight

JUNIOR HIGH - HIGH SCHOOL

▶ Explain how a hot-air balloon works.

▶ Explain how jet propulsion works.
Which of Newton's laws of motion is part of your explanation?
(*principle of action and reaction*)

▶ Describe the forces involved in flight: lift, weight, thrust, and drag.

▶ Define updraft and downdraft and explain their role in flight.

▶ How does a propeller provide thrust and lift for an airplane?

▶ How do ailerons, elevators, and rudder help control an airplane in flight?

▶ Explain Bernoulli's principle and how it applies to flight:

If a fluid is moving along a surface it will cause a force on that surface which is perpendicular to that surface. If the fluid moves faster, it will cause less force on the surface.

  A study of spacecraft and rockets could be included here.

In the blank under the student's name, fill in the abbreviation for each skill level as it is completed:
P *(primary)*   **I** *(intermediate)*   **Jr.** *(junior high)*   **Sr.** *(senior high)*

CHECKLIST	NAME				
Scientific Skills					
Observe					
Classify					
Measure					
Predict					
Infer					
Control Variables					
Interpret Data					
Formulate an Hypothesis					
Define Operationally					
Experiment					
Research					
LIFE SCIENCE					
ANIMALS					
Living / Nonliving					
Distinguishing Characteristics					
Classifications					
Life Cycles					
Habitats					
Ecosystems / Biomes					
CYCLES					
Water					
Carbon Dioxide - Oxygen					
Nitrogen					
PLANTS					
Living / Nonliving					
Distinguishing Characteristics					
Habitats					
Uses					
Growth					
Reproduction					
Response to Stimuli					
HUMAN BODY					
Parts of the Body					
Five Senses					
Ears and Sound					
Eye, Sight, Light					
Teeth					
Cell					
Skin					
Blood / Circulatory System					
Digestive and Excretory Systems					
Endocrine System					
Skeletal and Muscular Systems					
Nervous System					
Reproductive System					

CHECKLIST	NAME				
HUMAN BODY continued					
Respiratory System					
Care of the Body					
Causes, Prevention, Treatment of Illness					
EARTH SCIENCE					
Weather					
Weather and climate					
Atmosphere					
Solar system					
Water					
The Land					
PHYSICAL SCIENCE					
Motion					
Gravity					
Work / Power					
Work / Machines					
Simple Machine: Wheel and Axle					
Simple Machine: Inclined Plane					
Simple Machine: Levers					
Water as Force					
Air as Force					
Magnetism					
Heat energy					
Radiant energy					
Electricity					
Waves					
CHEMISTRY					
Matter					
Atoms, Elements, and Compounds					
Gases					
Water					
Solids: Metals, Nonmetals, Crystals					
TECHNOLOGY					
Electronics					
Engineering					
Flight					